SAFETY IN SCIENCE EDUCATION

London: The Stationery Office

Department for Education and Employment
Sanctuary Buildings
Great Smith Street
London SW1P 3BT
0171 925 5000

ISBN 0 11 270915 X
Third Impression 1999

Contents

Functions

1 Introduction

As with any practical activity there is an element of risk in school science. However, this can be kept to an acceptable minimum if those involved are aware of the potential hazards and take appropriate steps to avoid accidents. Safety legislation places the main duty of care on employers and it is they who have the main responsibility for working out safe systems and delegating functions to teachers.

This guidance is aimed at secondary schools and colleges in England and Wales teaching science at Key Stages 3 and 4 and GCE or GNVQ, A-level or their equivalent. Nevertheless, it will have some relevance for similar situations in Scotland and Northern Ireland, and for middle schools and special schools. This guidance supersedes previous safety guidance in science issued by the Department.

This guidance is intended to help schools by setting out clearly and comprehensively the statutory requirements to which they must have regard. It also sets out good practice guidance which schools may adopt in order to comply with their legal obligations, having regard to the best use of resources available to them. The guidance is in no way intended to inhibit the teaching of practical science. School science is safe. Few accidents occur in science laboratories in comparison to other activities undertaken in a normal school day. When accidents do occur, they are usually minor. Only a tiny handful result in serious injury. Nevertheless, almost all accidents could have been anticipated.

Primary responsibility for the school's health and safety policy rests with the employer, although it is important that school staff also clearly recognise their roles and responsibilities. Under the Health and Safety at Work etc Act 1974 (HSWA), employers are required to do all that is practicable to ensure the health, safety and welfare at work of employees, and the health and safety of non-employees affected by their duties such as pupils and visitors. It is important that teachers are aware of their responsibilities regarding health and safety and ensure that pupils act safely, within acceptable bounds, at all times.

In county schools in England and Wales the national curriculum requires that science be taught in a practical and investigative manner. While some teachers feel that their freedom to carry out practical work has been constrained by recent safety legislation, this is often due to misapprehension. There are fewer constraints than many teachers believe. The local rules laid down by some employers go beyond the norms adopted elsewhere and some require more documentation than others.

This book will be useful to all professionals directly or indirectly involved with the provision of science education. These include:

- teachers;
- technicians;
- headteachers;

- other managers, governors and safety officers;
- education authorities; and
- teacher trainers, INSET providers and curriculum developers.

It will also be useful for OFSTED, LEA and health and safety inspectors, architects, publishers examination boards, the Schools Curriculum Assessment Authority (SCAA), trade unions and professional organisations.

Copyright on this book is waived to the extent that educational establishments may freely copy parts of the book for use within the institution that purchased it.

The Department is grateful for the advice and comments made on various drafts of the publication from the Health and Safety Executive, Office For Standards in Education (OFSTED) and Office of Her Majesty's Chief Inspectorate (OHMCI) Wales, many teachers, advisers and inspectors. Particular thanks go to the Association for Science Education,(ASE) and the Consortium of Local Education Authorities for the Provision of Science Services (CLEAPSS).

PART A

ORGANISATION AND MANAGEMENT OF SAFETY

2 THE LEGAL POSITION: ROLES AND RESPONSIBILITIES OF EMPLOYERS AND EMPLOYEES

2.1 Responsibilities of the employer

Responsibility for complying with safety legislation lies primarily with the employer. The Health and Safety at Work etc Act 1974 (HSWA), the basis of the legislative framework, sets out the general duties which employers have towards employees and non- employees.

The employer is normally the organisation with whom an employee has a contract of employment. In county schools the employer is the local education authority (LEA). In voluntary aided, grant maintained, most independent schools and city technology colleges the employer is the governing body. In some independent schools the employer may be the proprietor or trustees.

The HSWA requires employers to:

- provide safe and healthy working conditions for employees and others;
- provide information, instruction, and training about health and safety;
- prepare a written safety policy including the organisation and arrangements for bringing that policy into effect. The policy must be brought to the attention of employees.

The Management of Health and Safety at Work Regulations 1992, made under the HSWA, make explicit what employers are required to do to manage health and safety. They require employers to:

- assess the risks to health and safety to which employees and others are exposed. The significant findings of the assessments must be recorded;
- make arrangements for the effective planning, control, monitoring and review of safety measures which should be recorded;
- provide comprehensive and relevant information to employees on risks, specific safety measures, procedures to be taken in the event of serious and imminent danger, and the identity of those responsible for implementing such procedures;
- appoint a competent person, or persons, to assist the employer in complying with his legal duties.

An LEA must provide or ensure the provision of safety guidance for county schools.

2.1.1 Governors

As stated above, the primary legal duty to comply with health and safety legislation rests with the employer. In practice the employer would achieve this by delegating safety functions to individuals.

Governors of a county school must ensure that their school follows the safety guidance laid down by the LEA.

The key document, which will help governors of county schools deal with health and safety, is the LEA's Health and Safety Policy.

The organisational arrangements for implementing the policy should include those safety management functions delegated to the governors.

In all cases governors are responsible for ensuring that safety is promoted. Well-defined safety functions should be explicitly delegated to specific staff who should be given time and resources to carry out safety functions, buy equipment, attend appropriate training etc. Governors should require an annual report from the headteacher relating to health and safety issues.

Useful advice is given in the HSC publication The Responsibilities of School Governors for Health and Safety (see Useful Publications).

2.1.2 Employers' liability for the actions of employees: Insurance

In civil law, employers are liable for the negligence of their employees which results in injury to non-employees in the course of their normal duties. This is called 'vicarious' liability. It is not compulsory to have insurance cover against such an action, but most employers do.

It is possible for an employee to be named separately in an action, perhaps because the alleged negligence was during an activity which was not required by the employer. Not all employers insure for such liabilities and teachers are advised to obtain their own cover, perhaps through a professional organisation such as the Association for Science Education, or their trade union.

2.2 Responsibilities of the employee

Under the HSWA and its Regulations employees must:

- take reasonable care for their own and others' health and safety;
- co-operate with their employers over safety matters;
- carry out activities in accordance with training and instruction;
- inform the employer of any serious risks and failures in safety arrangements; and
- not interfere with or misuse items provided for health and safety.

2.2.1 The headteacher

As the manager of the school, many of the functions concerning health and safety will be delegated to the headteacher. Many headteachers will further delegate functions to other members of staff.

2.2.2 The head of science

As managers of their department, heads of science should expect to have safety functions relevant to the department delegated to them. They may, in turn, delegate functions but will need to ensure that delegated functions are carried out. It is important that delegated functions are clearly understood through job descriptions and through a description of the safety management structure within the school safety policy. Delegating a function or a task to another person does not necessarily delegate liability in the event of an accident.

In county schools it is likely that competent advice (as required by the Management of Health and Safety at Work Regulations 1992) will be available from the LEA. In schools where the governors are the employers they may appoint one of the teaching staff, for example, the head of science, as the competent person. Part of his/her duties may include monitoring new safety legislation. The employer should ensure that anyone who is appointed as a competent person is given adequate time, training, and resources to fulfil their function. This is of course the case for any person to whom a safety management function is delegated. Organisations such as CLEAPSS monitor new safety legislation on behalf of schools and membership provides easy access to up to date information.

The training required for staff to carry out a particular safety function will need to be assessed by the employer. The extent of the training will obviously depend on the initial level of competence of those staff. Advice on suitable training courses can be obtained from organisations such as ASE or CLEAPSS.

2.2.3   Other staff

Staff are expected to take reasonable care for their own and others' safety, in other words a care dictated by common sense. An example of failure to exercise this care might include directing pupils to make gunpowder.

Staff must co-operate with their employer, who has legal duties under health and safety law, by following their health and safety advice and instructions and by reporting unsafe conditions and practices. Staff expect to be given functions involving safety. A teacher might be asked to look through a course to check activities with hazards against general risk assessments; a technician might be asked to conduct a visual check on a mains cable for damage. Where functions involving safety are delegated to staff, the employer should ensure that such staff are given adequate time, training, and resources to fulfil their task competently. Otherwise they should carry them out as part of their duty to co-operate with the employer. All staff should know who has been given safety functions. In turn they should know to whom problems should be reported.

For technicians some specialised training may be needed, for example, in monitoring fume cupboards and portable mains-powered equipment.

Under common law teachers are deemed to be acting in 'loco parentis' which means they are expected to exercise the same degree of care which a reasonably prudent parent would do in the same circumstances. Under the terms of the School Teachers Pay and Conditions Document a teacher is under an obligation to maintain good order and discipline among pupils and to safeguard their health and safety.

Table 2.1 Key Management Health and Safety Functions

	Education Authority maintained sector	Independent, Grant-maintained, voluntary aided, CTCs, etc	Key functions in health and safety management
Employers	Education Authority or Education Dept	Governors, Governing body, proprietor	Commitment, monitoring resource management, performance reviews
Persons in control	Governors		
Senior managers	Chief Education Officers, Directors of Education, Education Officers, Headteachers	Headteachers	Commitment day-to-day management, arranging inspections, communication, resource management
Other senior managers	Deputy Headteachers, Bursars		
Other managers	Department Heads		Inspections, action, day-to-day management
Other employees	School Staff Teachers, Technicians, Caretaker/Janitor, Maintenance Staff, Cleaners, Administrators, Welfare Staff, Boarding House Staff, Assistants, Caterers, etc		Day-to-day management, participation in inspections, reporting defects
External advisers	Safety advisers, Consultants, Education officers, Property advisers, Maintenance advisers		Auditing, providing technical advice on standards and legal compliance

3 SAFETY MANAGEMENT IN THE SCIENCE DEPARTMENT

3.1 Legal requirements

Under the Management of Health and Safety at Work Regulations 1992, employers must ensure effective planning, organisation, control, monitoring and review of safety measures. These should be recorded. It is also a legal requirement under the Regulations to establish appropriate procedures for danger areas and emergencies and provide restrictions, information and training. Temporary employees must also have adequate information and training. Employers must have a documented safety policy.

Key elements of management in most science departments are the departmental staff meeting, record keeping, which will include instructions from employers (in some cases, quite detailed guidance or codes of practice) and heads of science, and, possibly, a diary or log. Departments will need some safety texts, including risk assessments.

Instructions and other papers to do with safety should be kept in a way that provides easy reference. Some of these documents can be considered as an appendix to the department's safety policy but others may be more conveniently kept separately in a dedicated safety file.

3.1.1 Health and Safety policy

Some employers may require specialist departments to have their own safety policy. It is sensible if the policy itself is short but supported by appendices that can easily be updated. Departments should play a part in devising their own policies.

One arrangement is to keep a reference copy of the policy in a ring binder with card dividers to separate the appendices. The binder can be kept with other safety texts in the preparation room or the departmental office.

Below is a suggested structure for a science department safety policy:

THE SAFETY POLICY OF THE SCIENCE DEPARTMENT OF XXXXXX SCHOOL

Status of the policy: it is a requirement of the employer that the policy must be followed by staff, that a copy must be kept in the department and that any amendments issued must be added to it.

1 Why this policy is needed: brief explanation.
2 General aims and duties: brief commitment by staff to promote health and safety followed by an outline of basic HSWA requirements.
3 Specific requirements:
   3.1 legal requirements, recommendations by the DfEE, HSE, etc
       including: risk assessments, manual handling, fume cupboard testing,

security, electrical testing, emergency procedures, radioactive sources, blood and cheek-cell sampling, pressure vessels, animals and plants in schools, equipment, AMs relating to science, personal protective equipment, reporting procedures;

3.2 school rules for staff and pupils;

3.3 specific restrictions, which might be imposed by the employer or the head of the science by delegated authority, or a recommendation by DfEE;

3.4 other safety advice and good practice, with a recommendation that those in certain texts are followed;

3.5 science department schedules and check lists for monitoring equipment, etc, isted here but kept in an appendix;

3.6 science department drills for emergency procedures and remedial measures.

4 Duties of the head of the science department, noting that any delegation should be well-defined, and that the head of science should check that staff follow the policy and that delegated functions are carried out, and that there is a system of management.

Appendices

1 An index suggesting where further information can be found.

2 Specific laboratory safety restrictions and advice, such as DfEE Administrative Memoranda and circulars from the employer.

3 List of safety texts required to be kept in the department for reference.

4 Notes and schedules for the examination and testing of equipment including fume cupboards, portable mains-operated equipment, autoclaves and steam engines, gas cylinder regulators, together with a log of completed schedules and tests.

5 List of staff to whom functions have been delegated.

6 Remedial measures for science staff to carry out while waiting for first aid.

7 School injury reporting procedure.

8 Departmental rules for pupils, teachers and technicians, with checklists for daily, weekly, termly and annual monitoring and lists of equipment for which training is needed before use.

9 Management procedures, with an outline of the organisation.

3.2 Organisation

Safety organisation should be part of the general organisation of the department. Procedures should be put in place to ensure that the health and safety policy is implemented. These may include the following:

- items at departmental meetings;
- periodic checks;
- precise delegation;
- a system of risk assessments;
- a system of reporting accidents and near-accidents;
- budgeting and spending for safety equipment;
- training; and
- a system for circulating safety instructions and storing them for reference.

3.2.1 Communication

Science department staff meetings play an important part in communicating the safety requirements of an education authority or the head of the department. Technicians should be present when safety matters are discussed. If those teaching science part-time cannot be present, they should be informed quickly of what was discussed. Minutes or some simple record of discussion should be taken. Accidents and near-accidents should be discussed as a matter of course.

3.2.2 Control

Good control can be achieved by setting clear objectives and allocating responsibilities. The responsibilities of managers, teachers and other staff should be defined clearly in the health and safety policy. Heads of science departments or those to whom they delegate safety functions should not assume that, because an instruction is issued, it will be followed. It is important to make enquiries at staff meetings, be alert to the state of equipment left out in a laboratory, listen to comments by technicians and pupils, and check that staff observe risk assessments and that pupils follow safety instructions.

3.2.3 Planning

Risks can be minimised by sound planning. Employers need to identify hazards, assess risks, assign priority to the risks and decide whether further preventative action is necessary. Planning for safety in the Science Department could include:

- the phased replacement of equipment no longer considered sufficiently safe;
- a review of risk assessments in a particular area of the curriculum; and
- modifying in-house training in remedial measures to make them more realistic.

To be effective, plans need to be discussed at departmental meetings. This should involve other staff such as technicians. The minutes of the meeting may provide a useful record. Plans should contain realistic deadlines.

3.2.4 Monitoring and review

The management of Health and Safety at Work Regulations require employers to have arrangements in place to control health and safety. These need to include monitoring arrangements. The monitoring system checks that the management system is working and that risk control measures are effective and are being maintained. Enquiries at a departmental meeting and scanning minutes of past meetings will often provide information for a review; as will accident reports and the reports from systematic checks of equipment and premises. An annual one-page report to the headteacher can be used as an opportunity for a more formal review.

3.3 Diary

A laboratory diary or log book can be useful for, among other things, recording safety matters such as the date of a maintenance check or when a warning was issued.

3.4 Record Keeping

The precise arrangement will vary from school to school but the following may be included:

1 the safety policy with appendices;

2 minutes of departmental meetings;

3 reports on accidents, near-accidents, defects of the accommodation etc;

4 safety memos to the staff and to senior management;

5 addresses of useful organisations (CLEAPSS, ASE, etc);

6 reports by inspectors;

7 safety instructions for equipment;

8 annual reports to senior management or the governors;

9 diaries or logs recording details of informal checks on equipment; and

10 a list of safety texts used for risk assessments; note that these must be out on a shelf in the preparation room and not filed away.

3.5 Instructions from employers

Staff are obliged to follow instructions from an employer or the person to whom the employer has delegated the job of promoting safety. It is important that all staff, including part-timers, temporary and newly appointed staff, receive such instructions and it is prudent to record that they have received them.

In some schools, a copy of an instruction is circulated to staff and initialled to indicate that it has been read. In others, copies are issued at departmental meetings, the minutes recording those present. A copy of such instructions should be kept in the departmental safety file.

If an instruction is issued informally it is sensible to refer to it again at a departmental meeting so that it can be officially documented.

3.6 Safety rules for staff

There will always be informal staff rules; writing them on paper and calling them rules might seem to diminish professional status but, in fact, doing so often gives reassurance. Any rules drawn up by the science department should be endorsed by senior management in the school.

Staff rules should be brief and might include:

- the need to observe the department's health and safety policy, including its risk assessments;
- not leaving pupils unsupervised;
- dealing with disciplinary problems;
- carrying out security measures and checks before leaving science rooms;
- not eating, drinking or smoking in laboratories; and
- not conducting hazardous operations when alone.

3.7 New, temporary and part-time staff

It is important that these are given safety training appropriate for the work they are required to do. Even non-science teachers supervising a class doing book work in a laboratory must be warned of any hazards and of emergency procedures.

3.8 Personal protective equipment in the Science Department

Activities in science departments frequently require addition measures to be taken to avoid accidents to both staff and pupils.

Employers are obliged, under the Personal Protective Equipment at Work Regulations 1992, to ensure that:

- suitable personal protective equipment (PPE) is provided for employees;
- risks are assessed to determine what PPE is suitable;
- such equipment is stored and maintained appropriately;
- staff have sufficient information and instruction about what PPE to use: how to use it and how to maintain it; and

- the equipment is properly used.

In turn, staff must use the equipment in accordance with the instructions and must report any loss or defect in equipment.

3.9   Health and safety audit

An audit involves a structured assessment of the efficiency, effectiveness and reliability of safety management in a science department. It is not legally required but can be very helpful in the review process. It is often more effective when carried out by someone not involved in the day to day management of the activity being audited, such as the school's safety officer. Heads of science departments themselves should monitor the detailed implementation of their department's policies.

An audit should not become so concerned with detail that it fails to question policy. Auditing can be based on a series of questionnaires covering aspects of health and safety.

4 RISK ASSESSMENTS

4.1 The difference between a hazard and a risk

- a hazard is something with the potential to cause harm to a person, or damage to property;
- a risk is the likelihood of a hazard causing harm in practice.

4.2 The legal position

The Management of Health and Safety at Work Regulations 1992 require employers to make suitable and sufficient assessments of the risk to health and safety to which their employees, and non-employees, which would include pupils, are exposed by their work activity. The significant findings of these assessments must be recorded and the assessments must be periodically reviewed.

Where other legislation such as The Control of Substances Hazardous to Health Regulations 1994 (COSHH) Regulations require assessments of specific hazards these need not be repeated, but the employer must ensure that all significant risks are covered.

4.3 When a risk assessment is needed

An assessment is needed for accommodation or equipment or any activity, such as a class practical exercise or teacher demonstration, where a hazard may be present.

4.4 Who should carry out risk assessments?

The duty to carry out risk assessments rests with the employer. However, in practice, the task may be delegated to a member of the science department. The employer should provide adequate training, time and resources for the assessments to be carried out. The HSE Booklet `Five Steps to Risk Assessment' provides useful guidance on how to carry out and record risk assessments in the workplace. The Approved Code of Practice on the Management of Health and Safety at Work Regulations 1992 also provides useful advice. Where regulations such as the Manual Handling Operations Regulations 1992 require assessment of specific risks, employers will need to ensure that those staff who are given the task of carrying out the assessment are competent to do so.

4.5 General assessments

General assessments (also called generic or model assessments) are found in published texts. In themselves, they are not sufficient because a risk assessment should depend at least partly on local conditions. A general assessment may therefore need to be adapted.

When general assessments cannot be found in published texts, employers must make arrangements to carry out a special risk assessment. In many cases, such activities are sufficiently similar to standard ones for a general risk assessment to be adapted. However, there are organisations such as CLEAPSS and Scottish Schools Equipment Research Centre (SSERC) (see Useful Addresses) which have the specialised expertise to provide risk assessments for their members for such activities. The SSERC book `Preparing COSHH Risk Assessments for Project Work in Schools' and the CLEAPSS Guide, `Risk Assessments for Science' (see Useful Publications) are relevant.

4.6   Making a risk assessment

Hazards must be identified and considered. Account should be taken of the size of the practical group, supervisory ratio, age, ability and special educational needs in relation to the accommodation, equipment and activity. When assessing risk ask yourself the following questions:

- How could persons be injured or their health damaged during the activity?
- Could the activity go wrong and thereby produce hazards?
- Has this activity a worthwhile educational aim?
- Can this aim be achieved by a safer activity?
- Can a less hazardous chemical, organism or procedure be used?
- Could risks be reduced by changing parameters such as voltages, pressures or temperatures or the quantities or concentrations of chemicals?

Appropriate safety equipment should be used. Examples of safety equipment include eye protection, safety screens and fume cupboards. In all cases, the assessed procedures must be followed.

Pupils need to be told what the hazards are and how to ensure risks are kept low and appropriate emergency procedures must be known by staff and, where necessary, by pupils. Risk assessments will also need to take into account the needs and safety of pupils with special educational needs, teachers, technicians, other staff such as cleaners and any visitors.

4.7   Recording risk assessments

It is a legal requirement to record the significant findings of risk assessments. HSE Inspectors may ask for evidence of how a particular science department has responded to general assessments printed in published texts.

While some employers require the completion of a printed form, this approach may have disadvantages - see below. However, most employers ask school staff to incorporate risk assessments into the materials normally used in teaching, such as schemes of work, lesson plans or work sheets.

4.7.1 Risk assessment forms

Completing assessment forms which have been well designed to fit school science activities fulfils legal requirements; they demonstrate that general assessments have been consulted and that thought has been given in the department about risk reduction.

However, it is difficult to justify the time taken, because these forms may do little once completed to remind staff of hazards and how risks can be reduced. Their completion may involve only a few staff and, because they are separate from the main texts used daily in science teaching, there is a risk that they will remain in a filing cabinet and not be consulted regularly. In addition, they are likely not to be modified when an activity is changed.

4.7.2 Risk assessments on texts used every day

One approach for recording risk assessments is to annotate documents which are used daily in the science department. The information will be derived from the general assessments stipulated by the employer but will need adaptation for the circumstances of a particular department and/or classes; it will include appropriate hazard warnings, substitutions, restrictions and precautions.

Texts which can be annotated include schemes of work, lesson plans, work sheets and text books. Annotation of these documents has advantages:

- because these risk assessments are on documents used daily to guide activities, there is a better chance that they will be read and followed;
- the assessments are more likely to be modified when the activities have to be modified;
- because this way of presenting risk assessments uses existing structures, they will fit more easily into science department thinking, be less demanding to accomplish and so more likely to be carried out; and
- it will be easier for employers or HSE inspectors to relate risk assessments in this form to actual practice because they will be part of the texts used for guidance and out on the bench with the equipment.

This approach is equally suitable for non-chemical hazards. It encourages the view that safety depends on the design of an activity, on the equipment, quantities, timing and so on, and is not something to be considered later. It refers staff back to general assessments which have been made by specialists who understand the significance of exposure limits, voltage limits and risks associated with micro-organisms.

4.8 Areas in which risk assessments are likely to be necessary

4.8.1 General

These will include:

- flames;
- hot liquids;

- hot objects, for example tripods;
- sharp instruments;
- voltages above 25 V;
- lifting and carrying heavy objects; and
- using tools.

4.8.2 Activities requiring personal protective equipment

Risk assessments are needed to decide when the following are needed and which types of each are appropriate:

- eye protection;
- gloves; and
- safety screens.

4.8.3 Biology

These will include activities involving:

- chemicals (see below);
- living or once-living materials, including animals (particularly insects, birds and mammals), plants that could be poisonous or produce sensitisation, micro-organisms and material from butchers or abattoirs;
- field work and other out-of-school activities;
- human body fluids including saliva, urine, blood (the direct sampling of blood is not recommended by the DfEE and is forbidden by most employers);
- taste testing; and
- hazardous equipment including sphygmomanometers and spirometers.

4.8.4 Chemistry

Risk assessments are required for activities involving chemicals which are:

- very toxic;
- toxic;
- harmful;
- corrosive;
- irritant; or
- with a Maximum Exposure Limit (MEL) or an Occupational Exposure Standard (OES). Assessments will also probably be needed for activities using:
- highly flammables;
- flammables; or
- strong oxidisers.

Other activities also likely to need risk assessment include those involving:

- electrophoresis with voltages over 25 V;
- exothermic reactions; or
- the generation of gases in closed vessels.

4.8.5   Physics

Risk assessments are required for activities involving:

- chemicals;
- electron and gas-discharge tubes, for example some Teltron tubes because of the use of high-tension units;
- ionising radiations;
- large masses;
- lasers;
- lifting beams and hoists;
- high pressures or vacuums;
- steam engines;
- stroboscopes;
- viewing the sun; or
- wires and plastic monofilaments under tension.

4.8.6   Technician operations

Technician operations likely to require risk assessment include those involving:

- lifting;
- carrying;
- pushing trolleys up ramps;
- animals;
- chemicals; or
- microbiological cultures.

Attention can be drawn to many of the risk assessments which technicians must observe by well-sited labels, for example, on heavy equipment which requires two persons to carry it.

4.9   Monitoring Health and Safety

The Management of Health and Safety at Work Regulations 1992 place a duty on employers to monitor safety and they should have written arrangements for carrying this out. The school's safety policy should outline the safety management functions of the headteacher and other staff and should detail exactly what level and type of monitoring is expected of them. Monitoring health and safety arrangements gives information for putting things right and, in the longer term, for reviewing policy and for planning and organising risk control.

5 EMERGENCY PROCEDURES

Risk assessments should not only cover hazards normally associated with an activity but those which may arise if the activity goes wrong. Science staff should be aware of such hazards and be competent to take the necessary action in the event of imminent danger to pupils, colleagues or themselves. Disabled pupils may need particular but discreet consideration; for example, exit routes may need to be planned for any wheelchair user.

5.1 Fire

Emergency exits must be adequately labelled and keptclear. Fire doors should not be wedged permanently open.

5.1.1 Fire and security alerts

Staff must be familiar with the procedures to follow if the alarm sounds indicating a fire or a security alert. These include:

- switching off the gas supply in the laboratory;

- shutting windows and doors;

- securing any hazardous items or materials such as certain chemicals or radioactive sources; and

- conducting pupils along the designated route to the designated area outside.

The science department's safety policy should summarise any procedures additional to the school's general list. All staff should know how to extinguish lighted hair or clothing (see section 6 First Aid) and some staff should be trained to use fire extinguishers.

5.1.2 Bench fires

All staff handling highly flammable liquids, including methylated spirit, should be trained to deal with small bench fires. Evacuation of persons must have priority but, if the fire is very small, the following actions should be taken:

- Bunsen burners and/or other flames should be extinguished;

- if liquid is burning in a container such as a beaker, it should be left to burn away or smothered with a heat-proof mat or a fire blanket;

- spills on the bench or floor up to 30 cm by 30 cm in area can be smothered with a fire blanket; and

- then a carbon dioxide fire extinguisher may be used to complete extinction, if necessary.

Extinguishers should be used with care, being turned on while directed away from the fire and then cautiously brought near it. It is easy to extend a fire by spreading the burning liquid or knocking over beakers of blazing liquid.

5.2 Cutting off gas and electricity

All staff should know how to cut off mains gas and electricity in any laboratory or preparation room they use. In some situations, plans showing the locations of cut-offs may be useful.

5.3 Fuming chemicals and gases

Staff directing work involving fuming chemicals and gases (for example chlorine, sulphur dioxide, ammonia and oxides of nitrogen) should check whether any pupils present are asthmatic, take precautions for these to avoid inhaling the fumes or gases and be able to take immediate remedial measures if they inadvertently do. In the event of an unintended significant escape of toxic gases and fumes, staff must first evacuate the site of the escape before taking steps to contain the release.

5.4 Spills

Every science department should have at least one chemical spills kit. Staff should be trained to use it. However, when spills of one or two litres of a fuming liquid such as concentrated ammonia or nitric acid occur in a confined space, such as a small store whose ventilation cannot be increased sufficiently, it may be necessary to call the fire brigade.

6 FIRST AID

6.1 The legal position

The HSWA 1974 places a general duty on employers to take reasonable steps to ensure that employees and other people on their premises are not exposed to risks to their health and safety.

The Health and Safety (First-aid) Regulations 1981 came into effect on 1 July 1982. The Regulations and their associated Approved Code of Practice apply to all employees. Although the Regulations do not apply to non-employees, including pupils, HSE recommend that employers make provision for them.

Every school should have a written statement of its policy on first-aid covering both employees and non-employees including pupils. The statement should be brought to the attention of all concerned and should include such information as:

- the names and locations of first-aid personnel
- locations of equipment
- special arrangements for dealing with accidents away from the establishment or outside normal hours
- liaison with ambulance services.

6.2 Immediate remedial measures in the Science Department

It is good practice for science staff to have received some form of basic first aid training. Following an accident in the science laboratory it may not be possible to locate a first aider immediately but the timing of the first aid may be crucial to limit the harm caused.

Steps which should be taken while waiting for a first aider include the following:

- **Chemical splashes in the eye:** immediately wash the eye under running water from a tap or eye-wash bottle for at least 10 minutes. The flow should be slow and eyelids should be held back. Afterwards, the casualty should be taken to hospital.
- **Chemical splashes on the skin:** wash the skin for 5 minutes or until all traces of the chemical have disappeared. Remove clothing as necessary. If the chemical adheres to the skin, wash gently with soap.
- **Chemicals in the mouth, perhaps swallowed:** do no more than wash out the casualty's mouth. After any treatment by the first-aider, the casualty should be taken to hospital.
- **Burns:** cool under gently running cold water until first aid arrives.
- **Inhalation of a toxic gas:** sit the casualty down indoors in an area free from fumes.

- Hair on fire: smother with a cloth.

- Clothing on fire: push the casualty to the ground, with the flames on top. Smother flames by preading a thick cloth or garment on top if necessary. A fire blanket is ideal if very close by.

- Electric shock: taking care for your own safety, break contact by switching off or pulling out the plug. If it is necessary to move the casualty clear of the source of the electricity, use a broom handle or wooden window pole or wear rubber gloves. If the casualty is unconscious, check that airways are clear.

- Severe cuts or wounds: the immediate priority is to stop excessive blood loss by applying pressure to the wound for example using a pad of cloth. Do not attempt to remove imbedded bodies before the first aider arrives.raise the wound as high as possible and, if blood loss continues or the casualty feels at all faint, lower him to the floor. Protect yourself and other people from contamination by the blood.

6.3   Reporting incidents

All staff must be familiar with the school's procedure for reporting and recording details of accidents. There is a legal requirement under the Reporting of Injuries,Diseases and Dangerous Occurrences Regulations 1985 ('RIDDOR') to report fatal accidents or major injuries to pupils to the Health and Safety Executive.

The reporting of all injuries should be organised for the whole school. All but the most trivial injuries must be adequately and rapidly reported. Not only will this help any investigation into the cause and reduce the chances of repetition, but a report can be useful for defence in the case of a court action. An accident book for employees is required under Social Service regulations. Under RIDDOR, employers must immediately report to the HSE:

- any death caused by accident, followed up within seven days by a written report on the correct form, as well as certain injuries, including fractures; amputations; serious burns; loss of sight to an eye; serious eye injuries; electric shock injuries involving loss of consciousness due to lack of oxygen; acute illness requiring medical attention or loss of consciousness as a result of absorption of any substance by inhalation, ingestion or through the skin; or any injury requiring hospitalisation;

- any injury to a member of staff causing them to be off work for more than three days;

- any dangerous occurrences including explosions, bursts or releases of chemicals which could have caused death or serious injury; or

- a work-related illness, such as an allergy to a certain substance, contracted by a member of staff.

In addition, employers must keep records of any event or disease which are required to be reported under the regulations.

6.4   First Aid and HIV

Qualified first aiders will be aware of the precautions which must be taken to avoid possible infection by the HIV virus, but it is essential that anyone administering aid to an injured person should also know what to do. There are standard precautions which reduce the risk of transmitting other infections, including hepatitis, and they are equally effective against HIV.

Exposed cuts and abrasions should always be covered up before giving treatment to an injured person and hands should be washed before and after applying dressings.

Whenever blood or other body fluids have to be mopped up, disposable plastic gloves and an apron should be worn; these items should then be placed in plastic bags and disposed of safely, preferably by burning. Clothing may be cleaned in an ordinary washing machine using the hot cycle.

The HIV virus is killed by household bleach and the area in which any spills have occurred should be disinfected using one part of bleach to ten parts of water; caution should be exercised as bleach is corrosive and can be harmful to the skin.

If direct contact with another person's blood or other body fluids occurs the area should be washed as soon as possible with ordinary soap and water. Clean cold water should be used if the lips, month, tongue, eyes or broken skin are affected, and medical advice should be sought.

Mouthpieces are available for first aiders giving mouth-to-mouth resuscitation, but they must only be used byproperly trained persons. Mouth-to-mouth resuscitation should never be withheld in an emergency because a mouthpiece is not available. No case of infection has been reported from any part of the world as a result of giving mouth-to-mouth resuscitation.

For further information see the booklet AIDS and Employment, published by the Department of Employment and the HSE.

7 WORKING WITH PUPILS

7.1 Safety education for pupils

Concern for safety within school science should not aim just to make school science laboratories safe work places; it should contribute to the safety education of pupils. As such, restrictions and precautions in the school laboratory should be explained as well as enforced. In carrying out risk assessments, consideration should be given to instructions, training and rules for pupils which relate to their safety.

It is advisable for pupils' work sheets and other texts to have safety warnings marked on them. However, teachers should support written warnings with oral warnings.

7.2 Safety rules for pupils

Every science department should have a set of rules for pupils. Each school should devise its own, and the set given below should be regarded only as an example. Pupils need to be reminded of the rules orally; reference to a printed set is not likely to be adequate. Rules should be:

- short, with specific extra instructions introduced as and when needed;
- enforced, if they are to command any respect; and
- clear, direct and as positive as possible.

An example of a set of safety rules for pupils follows.

SAFETY RULES

Normal rules for good behaviour in the school are even more important in the laboratory because of the particular hazards. As in other classrooms there must be no rushing about or fooling, and nothing must be taken away without permission. In science laboratories there are some additional rules:

1   Enter a laboratory only when a teacher says you can.

2   Use equipment or materials only when a teacher has given you permission. Follow the teacher's instructions carefully.

3   Wear eye protection when you are told to and keep it on until you are told to take it off when all practical work, including clearing away, is finished.

4   When you are told to use a bunsen burner, make sure that hair, cardigans, scarves, etc are tied back or tucked in to keep them well away from the flame.

5   When you are working with liquids, always stand up never sit. That way you can move out of the way quickly if something spills.

6   Never taste anything or put anything in your mouth when you are in the laboratory. This includes sweets, fingers and pencils. These objects might have picked up poisonous chemicals from the bench.

7   If chemicals get on your hands or on any other part of your body or on your clothes, wash them off. If you have been working with chemicals or with plant or animal materials, wash your hands afterwards.

8   Report any accident to the teacher. This includes getting chemicals in your mouth or eyes, or on your skin or clothes. Also report any burns or cuts. If you get a burn, cool it with plenty of water under a running tap. Report all breakages and damaged equipment.

9   Keep your bench clean and tidy. Put all bags and coats out of the way. Wipe up small splashes with a damp cloth and report bigger ones to the teacher.

10 Always put any waste solids in the correct bin, never in the sink.

7.3   Supervision of pupils

Pupils should be allowed in laboratories only if supervised by a teacher who is aware of the hazards and how they can be avoided. This restriction should appear in the science safety policy under both pupil and staff rules. Laboratories should not be used by pupils during recreation periods or even for registration by teachers who are not scientists.

7.4   Safe behaviour

Risk assessments assume that an aim of school science is disciplined good practice, although they should not assume that, with pupils, it will always be achieved. Disciplined good practice has general characteristics (a systematic approach to work with unauthorised activities strictly prohibited) and specific ones (for example, not overfilling test tubes and keeping them always pointed away from faces).

7.5   Indiscipline

Indiscipline on the part of pupils increases risks in practical activities and risk assessments should take it into account. It is very important that schools and science departments have explicit policies which support individual teachers by addressing problems caused by indiscipline. Indiscipline includes minor carelessness, such as prematurely removing eye protection. Habitual troublemakers may need to be excluded from certain practical classes or moved to a class taken by a more experienced teacher.

7.6   Class size

There is no statutory limitation on class size in any subject in schools in England and Wales. Teachers who are concerned that risks in practical work are increased to an unacceptable level because of the class size should report their concerns to the head of their science department and, if necessary, their headteacher. It may be possible to adopt alternative methods for particular pieces of work. However, if risks cannot be made acceptable, the activity must cease until it can be resumed safely.

7.7 Pupils with Special Educational Needs

School health and safety policies, including risk assessments, should take account of children with special educational needs. Pupils with learning or physical disabilities are entitled to be treated like all other pupils as far as practicable. However, additional safety measures may be required in the science department to take account of:

- the pupil's ability to understand instructions, follow them and understand any dangers involved;
- the pupil's ability to communicate any difficulty or discomfort;
- any physical disability which might affect the pupil's ability to perform a task safely;
- the scope for accommodating the needs of pupils with physical disabilities by adapting existing equipment or acquiring modified items for example laboratory benches which will allow a wheelchair to go underneath them. CLEAPSS can provide information on equipment for pupils with motor difficulties;
- any medical condition which may be adversely affected by exposure to equipment or materials.

Care should be taken to ensure that children are not put at risk because they have limited understanding of safe practice.

7.8 Pupils as subjects of investigations

Various investigations, in particular those studying aspects of human physiology, may involve work in which measurements of body function are made, for example during exercise. Precautions must be taken to ensure that there are no risks to the health and safety of the pupils. In addition, care should be taken to avoid emotional stress to pupils, for example during work on human genetics.

7.9 Visits and field work

For pupils working or travelling off-site, at least as much care for their safety must be taken as when they are within school. This can be achieved only if teachers leading groups of pupils have made adequate plans and prepared for the visit or field work. There must be an appropriate ratio of teachers to pupils. Special provision may need to be made with regard to such things as insurance and parental permission.

Employers' rules and guidance on off-site activities must be checked and complied with.

Hazards need to be identified in advance and precautions taken. Where a visit or field trip may involve some element of outdoor pursuits, it is essential that leaders of groups of pupils have received necessary training and hold appropriate qualifications.

7.10   Work experience

Education employers and organisers of work experience have the duty of ensuring that pupils are not exposed to risks to their health and safety when on work experience placements. It is a legal requirement that employers must make arrangements for work experience placements to be as safe as is reasonably practicable. They are likely to delegate the function of organising such placements to employees who should follow the guidelines and instructions which their employers must issue.

Employers have the responsibility for health and safety in the workplace. Current legislation also entails that pupils on work experience must be treated as if they were employees. Consequently employers are advised to ensure that pupils are properly briefed about health and safety issues, and that adequate supervision arrangements are in place, before a work experience placement begins.

In turn, the HSWA places a duty on employees, including pupils on work experience, to take reasonable care for the health and safety of themselves and anyone else who may be affected by their acts or omissions, and to co-operate as much as necessary with their employer and others to ensure that obligations imposed on the employer by health and safety legislation can be complied with.

Legislation allows pupils to take part in approved work experience schemes during the last 12 months of compulsory schooling (that is, after the Easter term in Year 10 and beyond). The legislation on work experience does not apply to pupils over the statutory school-leaving age, although education authorities and governing bodies will usually require that schemes for these students comply with existing guidelines for younger pupils.

Work experience is prohibited where placements are subject to statutory restrictions based on age limits, including working with various dangerous machines, with lead and on ships. In addition, the restrictions of locally applicable bylaws must be observed (see 'Work Experience A Guide for Schools' and 'Work Experience A Guide for Employers' under Useful Publications).

8 RESOURCES: EQUIPMENT, MATERIALS AND SPECIMENS

8.1 Buying resources

Employers have a duty under the Provision and Use of Work Equipment Regulation 1992 to ensure that equipment is safe and suitable for its intended use. In practice responsibility for purchasing science equipment may be delegated to science staff or technicians. The employer is, however, responsible for ensuring their competence to assess the suitability and safety of any resources purchased. Headteachers may be expected to monitor purchasing and contracting procedures to ensure that their employer's health and safety policy is complied with.

Employers should be aware that whilst safety functions can be delegated, legal responsibility for safety rests with them.

Suppliers have an obligation to provide resources which are safe and/or are supplied fully documented with warnings or hazards and instructions for safe use.

Instructions with safety implications have a legal status. They should be photocopied immediately the equipment is received and the original kept in a safe place. It is sensible if the equipment or its immediate container is labelled to indicate the existence of safety instructions.

A number of factors will need to be considered when assessing the suitability of resources for purchase:

- **the nature of the resources:** experience and knowledge of safety texts will suggest items which need careful purchase but they include:

    biological material, living or dead;

    chemicals classified as hazardous;

    equipment involving voltages above 25 V, high or low pressures, high tensions, high temperatures or fast-moving parts;

    equipment for measurements on the body; and

    safety equipment.

- **the intended users:** before an item is purchased, the purchaser should be clear who is going to use it. Is it for staff or pupils? If for pupils, at which Key Stage?

- **what is available:** purchasers are expected to obtain the safest items, even if this involves paying rather more, so far as is reasonably practicable.

- **the supplier:** a well-established supplier of school science products is likely to provide resources suitable for the purpose advertised. The suitability of items from general education suppliers who sell a limited range of science resources and from suppliers who cater for the general laboratory market is less certain. Those selling resources for DIY, for home electronics, for pets and for medical specialists do not usually have schools' use in mind when designing or selecting their products.

- **the existence of a European Community product directive:** before long, certain items used in laboratories, other than school laboratories, for example waterbaths, will need to comply with EC product directives and will carry a 'CE' mark. Those buying an item with or without the 'CE' mark must ensure that it meets essential safety requirements.

Existing equipment does not have to meet current standards until 1997. However, if there is a serious hazard, use of such equipment must be discontinued immediately. Secondhand equipment must be assessed very carefully before purchase, as must gifts before acceptance.

8.1.1 Centrifuges

Any purchased now must conform to BSEN 61010-2-020. Existing centrifuges not conforming to the safety requirements of this standard should not be used after 1 January 1997.

8.1.2 Radioactive materials

Possession and use of sealed sources, uranium salts, etc. by county and GM schools is controlled by the former DFE and WO AM 1/92 (see Useful Publications). County schools (through their education authority), GM schools and CTCs must apply to the D*f*EE for approval to purchase and use radioactive sources. Approval must be obtained before orders are placed with suppliers. Incorporated colleges and independent schools must inform HM Inspectorate of Pollution if they acquire radioactive materials (see CLEAPSS Laboratory Handbook).

8.1.3 Pharmaceutical products and drugs

Schools will receive enquiries from suppliers or even the police if they order certain substances, for example phenylacetic acid. This is a formality required whenever certain substances which can be used to make drugs, are purchased. All that is required is an assurance that they will be used for normal laboratory exercises.

8.1.4 Alcohol (ethanol)

A letter of authorisation from HM Customs is required before industrial methylated spirit can be purchased. Mineralised (blue) spirit does not require one.

8.2 Using resources

It is a legal requirement that resources purchased for a workplace must be used safely. This can often be achieved by general training in good practice, the application of risk assessments and appropriate restrictions and precautions.

In other cases, hazards are associated with the use of an item of equipment or a particular group of materials or organisms. For these, it may be easier to restrict the users, perhaps to staff, or to sixth form pupils under close supervision. Often safe use will depend on instruction and only those who have received this should be allowed to use certain items.

When the use of certain items is restricted to particular users, it is important that such items are kept secure and/or labelled so that the restriction is obvious. Standardised adhesive labels could be used for this.

8.2.1 Radioactive materials

Each employer needs a radiation protection adviser and each workplace a radiation protection supervisor (in practice this is often a physics teacher). Establishments need a short set of local rules. Pupils using radioactive materials, except some specified, very weak materials, must be aged over 16. For basic (Class C) work with sealed sources, qualified science teachers do not require special training.

8.2.2 Pharmaceutical products and drugs

It is illegal to isolate certain drugs, for example codeine. Pupil investigations involving substances in which this could occur must not be permitted.

8.3 Resources requiring some instruction before use

8.3.1 Gas cylinders

These should be kept in a cylinder trolley or clamped to a bench. When not in use they should be kept in the same place, known to the caretaker. Staff should practise turning a cylinder on and off before demonstrating in front of the class.

8.3.2 Mains operated equipment

All users should be asked to look out for damage to plugs, cables, etc.

8.3.3 High pressure equipment

Check suppliers' instructions before use. Safety valves should be checked.

8.3.4 High voltage equipment

High tension units and electrophoresis apparatus with output voltages of over 25 V and currents of more than 5 mA can give fatal shocks. They should be kept secure and labelled to prevent unauthorised use. 4 mm plugs with retractable shrouds should be used with these units. The power line demonstration must be conducted on a low voltage or else the apparatus must be suitably insulated and used only be staff who have received instructions. Equipment, for example a demountable transformer, imported from abroad by an unfamiliar supplier, needs careful examination. In all cases, the use of certain secondary coils giving more than 25 V should be restricted (see CLEAPSS Laboratory Handbook and ASE Topics in Safety).

8.3.5 Human body measuring equipment

This should be labelled to warn of the need to consult suppliers' instructions.

8.3.6 Powered and unpowered tools

Users must have sufficient instruction before tools are used. Staff who are unfamiliar with the use of tools and with teaching pupils to use them must take instruction from someone with appropriate experience. BS4163 should be followed, and training for some power tools resulting in certification is required by most employers.

8.4 Moving and transporting resources

The Manual Handling Operations Regulations (1992) concern the lifting or moving of heavy loads, and affect both employers and employees. Employers are required to take appropriate steps to reduce the risk involved in manual handling and to provide employees with appropriate safety information.

8.4.1 Within the laboratory

It is a legal requirement that, to protect the handler, manual handling which involves risk should be avoided. In a science department, the risks are most likely to arise from the careless handling of moderate loads. Trolleys should be used for heavy items and care taken when these items are lifted on or off. Storage should be arranged to minimise lifting. If heavy items, for example low voltage units, have to be moved frequently between different sites or up and down stairs, consideration should be given to the purchase of another set.

8.4.2 Heavy items

Before handling heavy items, for example carrying a vacuum pump upstairs, risks must be assessed. This might require precautions such as the use of two people (see CLEAPSS Laboratory Handbook). Trays and stacking boxes are a convenient way of carrying equipment. They should not be so full that they are too heavy to handle.

8.4.3 Corrosive substances

A 2.5 litre bottle of acid or alkali should be lifted with two hands, one under the base. If carried further than across the room, a carrier should be used.

8.4.4 Transporting chemicals between schools and sites

Science staff with laboratories on two sites are advised to plan carefully to reduce the need to transport chemicals by car or school van between the sites. If they have to do so, they must ensure that the packaging (the bottle, cap or stopper, packing, surrounding carton, etc.) are such that the contents are contained in the event of an accident and that the outer package is labelled so that those dealing with the accident have adequate warning. Most chemicals, in the quantities in which schools are likely to want to transport them, will be exempt from specific packaging and abelling requirements. However, to be sure of meeting them, it is advised that the most toxic, corrosive and highly flammable substances are transported in the complete labelled packaging in which they were supplied (see CLEAPSS Laboratory Handbook).

8.5 Storing resources

Heavy items should be stored such that their handling is minimised. Ideally they should be kept at a level at which they do not have to be lifted up or down a significant height; this reduces the risk of injuries to backs and of objects falling on the handler. If a heavy object is used so seldom that it has to be stored up high, then, before it is lifted down, a risk assessment must be made.

Long objects should never be stored so that they protrude into any walkway. Particular care should be taken if the ends of the objects are sharp.

If glass rods and tubing are stored on shelves, these should be designed that the rods or tubes cannot roll off. Such glassware should never protrude into a walkway.

As far as possible, books and papers should be stored away from chemicals, flammable liquids and oxidizing agents.

Ordinary refrigerators or deep freezes should not be used for storing flammable liquids or anything giving off flammable vapours (including some biological preparations) as there is a risk of the automatic light switch or thermostat igniting the vapour-air mixture. A refrigerator used for storing biological materials and chemicals should not be used for anything which is to be eaten, drunk or tasted.

8.6 Inspecting and maintaining resources

8.6.1 Checking and maintenance

The HSWA places a general duty on the employer to provide and maintain equipment which is, so far as is reasonably practicable, safe and without risks to health or safety. Some items are subject to specific regulations, for example PPE which stipulate that they must be "properly maintained".

The frequency at which equipment should be checked for safety is sometimes specified by law. However, some equipment is not required by regulation to be checked regularly but expert guidance suggests that this is advisable:

- portable mains-operated electrical equipment should be checked regularly with a tester; whether testing is annual or less frequent (for example only every three or five years) depends largely on how often the equipment is used or moved;

- mechanical ventilation (for example an extractor fan) which is needed by a risk assessment should be checked every 14 months to ensure that the average airflow has not fallen by more than 10 per cent. A quick check should be made each time the system is switched on to make sure it is extracting; and

- high pressure systems (for example autoclaves and steam engines) need regular visual inspection to check that rubber parts are not perishing, valves are functioning, etc. Written instructions on the safe use of the equipment must be available to users.

8.6.2 Maintenance without formal checking

Some equipment must be maintained to keep it safe, but no formal inspection is specified. Schools are advised to draw up their own lists of these items and decide how to ensure proper maintenance. Much depends on how frequently equipment is used. Some items, such as tools in a workshop, might be checked once a term. Others should be checked before each use, a simple notice to this effect being attached to the box or tray in which they are stored. For example, a hoist designed for car engines but used to teach about machines needs on its container a label that reads: 'Before lifting a load of more than one or two kilograms, check that the cord shows no signs of wear and that each end is firmly attached. Make sure the hoist is properly supported'.

In addition, there is a need to inspect workplaces for safety at regular intervals, such as every evening, to ensure hazardous items are secure and all gas taps are completely off.

Where there is no legal or national guidance, the employer should specify, or at least give guidance, on the frequency of inspections.

8.6.3 Who can check equipment?

Most checks and maintenance in the school science laboratory (with the exception of the five yearly check on cylinder regulators) can be conducted by school staff. Staff who have had a short training session can be regarded as competent for basic checks required by regulations or official advice. However, they should know where to turn if they have any doubts. Bodies such as CLEAPSS and SSERC can give help to their members over the telephone. There is, however, occasionally the need either for a visit by a specialist in the case of large items, such as a fume cupboard, or to take a portable item of equipment, perhaps because it is giving dubious readings in electrical tests, to a local centre or contractor.

8.6.4 Records

It is important that records are kept for checks specified by regulations or official advice. As well as recording the information required by the appropriate schedule, they should state clearly whether an item is safe to use. In some cases, the records for several years can be contained on a single sheet, for example one for each fume cupboard. However, the large number of portable mains-operated items makes it sensible to record results on index cards or a computer database.

If records have to be sent to the education authority, copies should be kept in the establishment.

9 STORING CHEMICALS

9.1 Classifying chemicals for storage

It is important to classify chemicals so that they can be stored safely and found easily. A common mistake is to store together all potentially hazardous chemicals, such as those classified as toxic, corrosive and highly flammable. It is in fact illegal to store flammables with other hazardous chemicals. Moreover, storing corrosive volatiles in a metal cupboard designed for flammables will soon reduce it to rust.

A recommended classification for storage purposes is given in the table at the end of the section. Not all groups need completely separate storage but they may each need separate consideration in some circumstances. The groups are arranged here in order of priority, so that a chemical with attributes that would put it in more than one group is assigned to the group highest in the list. A toxic, corrosive, flammable liquid is thus listed under flammables.

Lists of chemicals which schools might have under each storage classification are given in the CLEAPSS Laboratory Handbook.

9.2 Highly flammable liquids

It is permissible to store up to 50 litres of highly flammable liquids in a laboratory, a preparation room or a store used for other chemicals in a metal cabinet meeting certain criteria, including an appropriate label. Such cabinets sold for the purpose by the normal suppliers or by fire equipment specialists will meet these criteria. Large departments may require two such cupboards which should be in different rooms some distance apart.

It is also permissible to have out for use in a laboratory or preparation room a few bottles, capacity 500 cm^3 or less, of highly flammable liquids.

Staff should know how to deal with spills, and there must be no sources of ignition present where dangerous concentrations of vapour might be expected. There must be no smoking in any place where a highly flammable liquid is present.

9.3 Care of chemicals in storage

Chemicals which have a short shelf-life, particularly those which can become hazardous, should be checked regularly. Bottles containing white phosphorus should be checked to see that the water level is adequate. Sodium and potassium are now likely to be supplied under liquid paraffin; unlike the naphtha used in the past, this does not evaporate significantly and so its level does not need checking.

The tops of bottles around the stopper, and the trays in which bottles stand, should be checked regularly and wiped with a cloth so that they are kept clean. If this is done and care is taken with the storage of corrosive volatiles, the smell associated with the storage of chemicals will be diminished considerably.

9.4   Labelling and warning signs

Any bottles or flasks containing a chemical should be labelled with the contents and the hazard symbol if the chemical is classified. Suppliers' catalogues usually indicate the classification, if any, of chemicals. Fire prevention officers often require a warning sign on the door of a cupboard containing radioactive materials and sometimes on the door of the room containing the cupboard. The sign should correspond to BS5378; those from reputable suppliers will.

Table 9.1 Grouping scheme of chemicals for storage

Category	Group/ Substance	Storage provision	Store away from
Special cases (do not fit into other groups)	Bromine	With corrosive volatiles. Close by should be 1M sodium thiosulphate (to neutralise splashes on skin) and anhydrous sodium carbonate (for spills).	Water-reactive solids
	Gas cylinders	1–3 small cylinders fixed to a bench etc. or kept in a trolley in an established place inside.	Flammable liquids
	Silicon tetrachloride	Bottle in a plastic box or desiccator, kept dry with silica gel.	Oxidisers and water-reactive solids
	Sulphur dioxide	Cylinder in a plastic box or polythene bag, kept dry with silica gel.	Corrosive volatiles and corrosive liquids
	Radioactive substances	Special cupboard or drawer, sited so that staff do not regularly sit or work near it.	Flammable liquids and corrosive liquids
	White phosphorus	Locked in an internal store or cupboard in the prep room, possibly with toxics. The water level should be checked regularly.	Oxidisers and water-reactive solids
Flammable substances	Flammable liquids, bulk	Fire-resisting store if more than 50 litres needs to be stored; otherwise a fire resisting cupboard.	Oxidisers and toxics
	Flammable liquids, working	Fire-resisting cupboard(s) in prep room(s). A cupboard must be provided in any prep room where substantial quantities are used (but contents must not exceed 50 litres).	Oxidisers and toxics
	Flammable solids and water-reactives	A locked chemical store. If the store is accessible to pupils, within a locked cupboard in the store or in a preparation room.	Oxidisers, corrosive liquids and flammable liquids
Toxic substances	Toxic chemicals	A locked chemical store. If the store is accessible to pupils, within a locked cupboard in the store or in a preparation room.	Flammable liquids
Corrosive substances	Liquids, acid	Chemical store or prep room at low level but protected from feet by a plinth or a rail.	Non-acid corrosive liquids, water-reactive solids and toxics
	Liquids, non-acid	Chemical store or prep room at low level but protected from feet by a plinth or a rail.	Acids, water-reactive solids and toxics.
	Corrosive volatiles	Internal cupboard with special ventilation or in a desiccator or a plastic box with silica gel.	Water-reactive solids
	Corrosive solids	Chemical store or prep room with general chemicals (inorganic or organic as case may be). Locked if there is pupil access.	
General chemicals	Oxidizers	Chemical store or prep room with inorganic chemicals. Locked if there is pupil access.	Flammable liquids, flammable solids, water-reactive solids, organics and corrosive liquids
	Other inorganics	Store or prep room, with labels on shelf to indicate location of chemicals in other groups.	
	Other organics	Store or prep room, with labels on shelf to indicate location of chemicals in other groups.	

10 FUME CUPBOARDS

10.1 Using a fume cupboard safely

Before using a fume cupboard, bottles and other items which are not needed for an operation should be removed. They can increase hazards if there is a fire or explosion. The ventilation should be checked. This can be done by attaching a strip of thin plastic film to the bottom of the sash at one end. Its deflection will indicate that the air is being drawn in.

It is important to respect the stop or mark that indicates the maximum sash opening. If the sash is lowered substantially from this position, enough room must be left for equipment to be manipulated and operated safely.

Operations in fume cupboards should be designed to avoid large, uncontrolled releases of toxic vapours. Proper eye protection should be worn - the sash should not be regarded as sufficient protection for eyes.

10.2 Checking fume cupboards: the legal position

A fume cupboard is an example of local exhaust ventilation equipment as defined by the COSHH Regulations, and as such must be thoroughly examined and tested every 14 months and records kept. The testing should include face velocity measurements. The examination should include visual inspection of ducts, sash cords, services, and so on.

10.3 Technical specifications

The recommendations of DFE Design Note 29 (DN 29 - see Useful Publications) need to be observed in the design and installation of new fume cupboards and, as far as possible, in upgradings.

10.3.1 Airflow:face velocities

When the fume cupboard sash is at its maximum height for normal work, the minimum air velocity into the opening below it (the face velocity) should be at least 0.3 m/s. Higher velocities reduce the effects of movements of the operator on the airflow but can make Bunsen flames unstable.

The variation in face velocity measured at the centres of nine cells across the opening should be below 30 per cent. This figure is important as a measure of stability of the air flow: if there is a greater variation, there is a greater tendency for eddies to occur which can bring the fumes out into the laboratory. However, an existing fume cupboard should not be failed if its variation in airflow exceeds 30 per cent.

10.3.2 Baffles and secondary air inlets

Baffles (panels at the back of a fume cupboard intended to help to achieve a uniform face velocity) are valuable, particularly if the fume cupboard is not very tall. Of more importance for ducted fume cupboards in which Bunsen burners will be used is the presence of a bypass or secondary air inlet. This comes into effect as the sash is lowered, usually by uncovering a vent at the top of the fume cupboard, and prevents the air flow under the sash from rising too much.

10.3.3 Maximum working height

The maximum sash opening should be fixed by a stop; a mark on the sash is insufficient. The stop should be capable of being unlocked with a key so that the sash can be raised further for cleaning inside or for assembling tall apparatus. DN 29 specifies that the maximum sash opening should be at least 400 mm. This is a compromise which gives sufficient room for most manipulations, offers some protection to the faces of operators and makes reasonable demands on the extraction system.

10.3.4 Minimum height

A stop fixing the minimum sash opening at about 50 mm is also important. Without it, a sash can be brought down to close the working opening completely, thus preventing the extraction of fumes.

10.3.5 Glazing

Some kind of protective glazing should be fitted. If an existing cupboard is glazed with ordinary glass, this should be replaced or covered with safety film.

10.3.6 Work surfaces

A melamine surfaced phenolic resin such as Trespa is the preferred working surface. Architects tend to favour light colours which soon become stained. Stainless steel can be used but can be pitted by spills containing concentrated chloride ions. Other suitable work surfaces include solid epoxy resin and quarry tiles with a grouting that is resistant to chemicals.

10.3.7 Double sashes

Fume cupboards fitted through the wall between a laboratory and a preparation room are unsafe: if they are fitted properly, with stops to limit the minimum sash opening, cross draughts will occur which cause vapours to escape.

10.3.8 Installation

The design of extraction systems requires a ventilation specialist familiar with DN 29: architects and builders on their own are inadequate. Centrifugal fans should be used in preference to axial fans, which are noisy. Discharge should be vertical and well above roof lines, particularly if these are pitched.

Fume cupboards will work effectively only if air can freely enter a laboratory to replace that which is being extracted. For this replacement air, vents to a corridor, for example, are desirable. These should be situated, if possible, opposite the fume cupboard; any lowering of the fire resistance of the wall should be avoided. The need for adequate replacement air becomes acute if several fume cupboards and/or extraction fans have to operate at the same time.

As well as the direction in which the replacement air will flow, other points to be considered when planning the site of fume cupboards are freedom from cross-draughts caused by doors and windows and from disturbance by those walking past.

10.3.9 Mobile fume cupboards

There are two kinds of mobile fume cupboard. Both are mounted on trolleys and can be moved. One kind is connected by a flexible duct to a fixed ventilation system. It's advantage is that it can be pulled out into the room and used for demonstration, with the teacher at the front and the pupils around the back and sides. In some installations, the flexible duct can be connected to the fixed system at various points, increasing mobility.

The other kind of mobile fume cupboard passes the air through filters before it is recirculated back into the laboratory. It has greater mobility and the absence of an external ventilation system reduces its capital cost considerably. However, filter-type fume cupboards have disadvantages: it is difficult to achieve satisfactory face velocity and adequate filter efficiency so that only a few models meet DN29 recommendations; the filter efficiency needs to be checked regularly as well as the face velocity; the cost of filter replacement which usually falls to the science department.

Whichever system of mobile fume cupboard is considered, it must be remembered that, if full services are to be provided, mobility is limited to the positions where the services can be connected safely through a cable and hoses to special outlets. Particular attention must be paid to the safe provision of gas and drainage.

11 DESIGN OF THE SCIENCE DEPARTMENT

11.1 Legal requirements

The principal piece of legislation concerned with the safe design of a science building is the Workplace (Health, Safety & Welfare) Regulations 1992 (see also DFE Building Bulletin 7, Useful Publications). Under these regulations, employers must ensure:

- adequate ventilation;
- that temperatures are reasonable;
- adequate lighting with as much natural light as possible;
- sufficient floor area;
- that work stations are suitable, with adequate seating;
- that when windows are open they cannot cause danger;
- that there can be free movement within the building; and
- that there are sufficient toilet and washing facilities and sources of drinking water.

Other considerations are:

- means of communication and evacuation routes in case of fire, injury and other emergencies;
- access for delivery, ambulances and fire appliances;
- security;
- movement of equipment and materials between stores, including any outside stores, preparation rooms and laboratories;
- the needs of wheelchair users; and
- storage for pupils' coats and bags.

11.2 Size of the laboratory

There are no regulations controlling the size of individual laboratories but adequate space is clearly needed for safe practical work. For 30 pupils at key stages 3 and 4, 85 m^2 is recommended; below 70 m^2 a laboratory will be appropriate for groups of 25 or fewer (see Science Accommodation in Secondary Schools, Useful Publications).

As well as overall area, the area of work surface available for each pupil affects safety; 0.36 m^2 is recommended as a minimum. Distances between work spaces are also important and useful guidance is given in Science Accommodation in Secondary Schools.

11.3 Sizes of preparation and storage rooms

A single, main preparation room should ideally have about 0.5 m^2 total floor area for each laboratory workplace. Thus, for example, if it is to serve five laboratories designed for a maximum of 30 pupils each, its area should be about 75 m². While the needs of different schools vary, a useful analysis is given in the table below.

Analysis of preparation room use

Use	Note	Percentage of area
Fixed storage	including chemicals	20-30
Mobile storage	including equipment trolleys	8-10 ≥
Working area	including space for sitting and standing	25-30 ≥
Circulation	for walking about, wheeling trolleys etc.	40-45 ≥

It is a good idea to have a room to be used as a staff base or departmental office. This is useful for taking refreshments, planning lessons and so on. It is convenient if such a staff base can be sited to allow observation of routes into laboratories which might be temporarily unlocked. If such a room is unavailable, in some preparation rooms it might be possible to allocate a 'chemicals-free' area for refreshments.

11.4   Security

Because science rooms contain equipment and materials which are hazardous if handled by untrained persons, security is important. The demands of security and emergency evacuation often conflict and need to be resolved.

11.5   Exits

Laboratories and preparation rooms should have two doors reasonably far apart so that evacuation is possible even if a fire or major spill emitting toxic vapour occurs near one of them. There should be more than one evacuation route from the room to the outside (DFE Building Bulletin 7, Useful Publications).

11.6   Fire doors

The positions of these within a building should be approved by a fire prevention officer. If possible, their positions should be such as not to hinder movement of apparatus.

11.7   Services and fittings

11.7.1   Mains electricity

The normal regulations governing the installation of mains electricity apply. In addition, each laboratory should have a master switch which enables the supply to all the bench sockets to be cut off. It should be clearly labelled, readily accessible to teachers and preferably close to other safety devices and equipment. There should be some sockets on the periphery of a laboratory which are unaffected by this switch, making them suitable for aquaria and other equipment which need a continuous supply.

The protection of the supply of each laboratory with a residual current device (earth-leakage circuit breaker) operating at no more than 30 mA and in less than 30 ms is recommended. Residual current devices protect only against leaks to earth and are no substitute for a master switch and proper maintenance.Sockets can be mounted close to sinks but should be positioned to minimise entry by water through splashing or overflow. Provision of outlets should be generous to avoid long leads or multiple adaptors.

11.7.2   Mains gas

The normal regulations governing the installation of mains gas apply. In addition, in each laboratory there should be a main valve which enables the supply to all the bench taps to be cut off (see British Gas/DES IM25 Guidance Notes on Gas Safety in Education Establishments; Useful Publications). It should be clearly labelled, readily accessible to teachers and preferably close to other safety devices and equipment. Because of a requirement that the main valve is also near the entry of the gas supply into the room, it is sometimes necessary to fit an electrically controlled valve at the entry point with an emergency switch or switches at accessible positions.

The design of gas taps on benches should be such that they cannot be turned on accidentally. Whether a tap is on or off should be apparent from across the laboratory. Gas taps should be mounted on or above the bench surface so that they are completely visible and towards the back of the bench to discourage fiddling. It is vital that tap mountings are properly fixed to the bench, with the smallest possible hole and slots cut for the antirotational lugs; fitters are often tempted to cut a bigger hole which goes right round the lugs. Those responsible for checking that work meets specifications should watch for this. Antirotational metal plates are now available and their use should be considered.

It is adequate for gas outlets intended for Bunsen burners to have corrugated nozzles on to which flexible tubing can be pushed.

11.7.3   Liquid petroleum gas

In schools where no mains gas is available, liquid petroleum gas (LPG) should be piped from cylinders in accordance with good practice (see Useful Publications). Because LPG is denser than air, it has a tendency to collect in low areas such as cupboards beneath benches. Consequently, leaks are more serious than for mains gas, which is less dense than air. The fixing recommendations given in the section on mains gas fittings are even more important.

11.7.4   Windows

The opening parts of laboratory windows should be well above the level of perimeter benches to avoid draughts affecting equipment such as Bunsen burners. It should be impossible for blinds to be blown into Bunsen flames. Excessive light makes Bunsen burner flames difficult to see.

11.7.5 Ventilation

The occasional release of fumes in a laboratory makes it desirable to be able to have five or six air changes per hour. While this can sometimes be obtained with natural ventilation through windows, these are likely to be closed in the winter or if the room needs to be dimmed for visual aids. Therefore, mechanical ventilation in the form of an extraction fan is desirable. However, this must be installed carefully so that it is not noisy and can still function when a room is darkened. It may be essential in an interior preparation room.

In any room fitted with mechanical extraction, there should be air inlets, particularly if there are also fume cupboards. If these are made in internal doors or walls, a fire prevention officer should be consulted, as there could be implications for fire safety.

Preparation rooms, particularly those in which chemicals are stored, often smell of chemicals, although concentrations of contaminants in the air are usually so well below occupational exposure levels that they are difficult to measure. While much can be done with good housekeeping, mechanical ventilation is sometimes necessary to achieve comfortable working conditions.

11.7.6 Smoke alarms

If these are fitted, it is important that they are automatically switched off when laboratories are in use as they can be triggered by many class activities involving heating.

11.7.7 Flooring

Laboratory floors are usually covered with industrial-quality thermoplastic sheet or tiles. Because laboratory floors can easily become wet, non-slip flooring is sometimes laid. If this is of the kind designed for swimming pool surrounds, namely incorporating abrasive particles, it will easily be marked by shoes and the caps on the bottom of laboratory stools; the caps can also be damaged. A less abrasive type of flooring is recommended, although this may be less effective in reducing slipping.

11.8 Storage

For chemicals, a separate purpose-built storage room is desirable. It should be sited so that it is not overheated by the sun. The safest and most convenient place is in the science department, as the transport of chemicals from external stores can produce hazards.

A store should have the following features:

- a dispensing bench;
- a concrete or quarry tiled floor sloped so that spills collect in an accessible

place where they can be removed with a spills kit;

- a low-level plinth for large bottles of corrosive liquids or at least a kicking rail; little or no heating (some may be necessary to prevent condensation on bottles, which loosens labels);
- an outward opening door with at least half-hour fire resistance, openable from the inside even when locked; and
- non-corrodible shelving.

Such a store can take all categories of chemicals, with the exception of highly flammable liquids. These may be kept in a flammables cupboard within the store provided their total volume is less than 50 litres.

Passive ventilation is most suitable for such a store, with vents in outside walls, some at high and some at low levels. If an extractor system with a fan has to be fitted, it should be a non-sparking type and acid proof. Lighting can be to normal standards. Lighting and fan switches should be outside.

12 INSPECTIONS

There are several categories of persons from outside a science department who might check on health and safety provision within it.

The following categories of inspector have the right to enter premises at any reasonable time and make enquiries concerning health and safety provision:

- **HSE Inspectors.** They enforce and advise on health and safety legislation in all educational establishments. Inspectors may visit as part of planned preventive inspection programmes, or to investigate accidents. Inspectors have the power to issue improvement notices, to remedy contraventions of health and safety legislation, or to issue prohibition notices to order the immediate cessation of activities which give rise to serious risk to health or safety. Serious breaches of legislation may result in prosecution of employers, or individuals. Prosecutions of teachers are however extremely rare, and usually result from a reckless disregard for safety.

- **A representative of a utility company,** such as water or electricity. They may enter premises if they believe bylaws are being disregarded.

- **HMI.** A specialist inspector may comment on safety management within the science department and draw attention to any deficiencies.

- **A science inspector attached to an OFSTED** team (England) or OHMCI team (Wales). They can comment on safety management, deficiencies or unsafe practices. They may only enter during a contracted inspection.

The following categories of inspector have a right of entry into county controlled schools:

- **Education authority safety officers.** They can issue safety instructions, ban certain procedures, and so on. They can investigate accidents on behalf of the education authority and initiate disciplinary action. They should advise on safety and in some cases can arrange for safety training of staff.

- **Education authority science advisers.** They can issue safety instructions, restrictions, etc, and can investigate accidents. They should advise on safety and may arrange training.

The following categories of inspector may enter a science department by arrangement with the employer:

- **A fire prevention officer.**

- **Inspectors of items such as fume cupboards, etc.** They may be acting on behalf of the school's insurers or a company with whom the school has a contract.

- **A school safety officer.** This is normally a teacher given the function of implementing safety arrangements. Usually expected to monitor safety arrangements and report problems to the headteacher.

- **A safety representative.** This will often be a teacher or technician elected by trade union members within a school and responsible to them.

PART B

SAFETY IN SCIENCE TEACHING

13 ABBREVIATIONS, SYMBOLS AND CONVENTIONS

For ease of reference, much of the information in Part B of this guidance is in the form of tables. The following recommendations provide guidance on what is considered good practice procedures. No activity should be carried out without first assessing the risks and taking account of local circumstances. To save space, the following conventions have been used.

Y7+, Y9+, Y12+	The chemical or procedure is generally considered suitable for use by pupils in England and Wales in Years 7, 9, 12 or above (Y12 implies GCE A-level) or their equivalent (that is, ages 11+, 13+, 16+). Note that local rules may apply a different, more stringent standard.
(Y7+), (Y9+), (Y12+)	While the activity/chemical is suitable for the year range indicated, stricter supervision than normal may be required, such as having a technician dispense the chemical or a teacher keeping direct control over a particular chemical. If this cannot be achieved, the activity should not be attempted.
L1, L2, L3	Categories of work in microbiology; they are not the levels of containment used by professional microbiologists.
T	The chemical/activity is suitable only for a demonstration or use by a technician or teacher.
X	The chemical/activity is banned nationally.
N	The chemical/activity is unsuitable or not recommended for use in schools. However, a small amount of chemical may be kept for exhibition purposes. In some areas these chemicals or activities may be banned by the employer's local rules.
F	A fume cupboard is normally necessary. Exceptions may be possible when using very small amounts of the substance.
(F)	A fume cupboard is desirable and should be used if available, but small quantities of the chemical can be used in a well-ventilated laboratory.

A complete list of acronyms and abbreviations can be found in Part C.

Because of their wide acceptance, the following abbreviated titles of publications are used: Hazcards, Topics, Safeguards, Haz Man, Micro, Handbook, Be Safe! Their full titles are given in Part C.

14 GOOD LABORATORY PRACTICE

14.1 Before the lesson

14.1.1 Lesson planning

Teachers should give technicians adequate notice of their requirements. Requisitions should not be left until the last minute, so that neither teachers nor technicians have time to check risk assessments. When planning lessons, allow sufficient time at the end of the lesson for equipment to be cleared away, benches to be wiped down, sinks left unblocked etc.

14.1.2 Emergency facilities

The location of fire extinguishers, emergency cut-offs, eye washing facilities and fire exits should be known to the teacher.

14.1.3 Distribution of equipment

This should be planned in advance, with several distribution points to avoid congestion and the risk of indiscipline.

14.1.4 Personal protection

Eye protection should be worn by pupils, teachers and technicians whenever the risk assessment for the activity requires it. Other protective or control equipment, such as safety screens or fume cupboards, should be used when required.

14.1.5 Tools

Tools may sometimes be used by pupils in science lessons. If so, there should be consultation with the Design Technology staff, to ensure that safety standards and procedures are comparable.

14.2 In the laboratory

14.2.1 Class control

Inattentiveness or poor behaviour leads to accidents. Staff should not feel inhibited about seeking help from more experienced colleagues.

14.2.2 Reminders

Pupils need to be reminded frequently of safe techniques, even when these should be familiar. Often a quick demonstration by the teacher will suffice. Do not allow pupils to crowd together, either for individual practical work or teacher demonstrations.

14.2.3 Coats and bags

These should be put well out of the way. Pupils will need frequent reminders. Ideally, there should be coat hooks in a quiet corner of the laboratory, and bags should not be allowed to clutter the floor.

14.2.4   Eating, drinking,chewing and smoking

These should not be permitted in the laboratory. If, exceptionally, a tasting activity is to take place, teachers must stress the special nature of this event and should adopt strategies to ensure that contamination cannot occur. Ideally, a home economics room or dining area should be used.

14.2.5   Hair

Long hair should be tied back and pupils should be warned that loose, flowing hair can make the hair more vunerable to catching alight.

14.2.6   Clothing

Check that pupils' clothing is suitable for the activity and does not, for example, present a fire hazard. Ties, scarves and cardigans should not be allowed to hang freely, as they could be a fire hazard or catch in machinery.

14.2.7   Electrical switches

These should never be operated by people with wet hands and pupils should be taught to switch off appliances before unplugging them. Pupils should be warned against meddling with switches. Electric cables should not be allowed to trail dangerously when being used, transported or stored.

14.2.8   Bunsen burners

When they are lit but not being used, Bunsen burners should be adjusted to show the yellow flame. They should be positioned carefully to avoid igniting wall fittings or blinds and so that pupils are not tempted to lean across them. Pupils should also be warned against meddling with gas taps.

14.2.9   Containers

Containers should be clearly labelled, with an appropriate name, any hazards identified (for example by a symbol) and the date of acquisition or preparation. Technicians or pupils sometimes fail to remove old labels, which can be confusing and hence dangerous. When containers are labelled, it is important to remember that the hazards of a solution are likely to be different from those of the substances from which it was made.

14.2.10   Extended practicals

Practical activities extended over a period of time should always be clearly labelled and dated, with any hazards identified. Equipment which is left running should have an appropriate warning notice (the one in Topics in Safety may be photocopied).

14.3   Using chemicals

The following guidance applies to all chemicals, and not just those used in chemistry lessons. A number of substances sometimes found in schools, for example Procion dyes, fumes from solder flux and some materials of biological origin, are sensitisers. Once exposed to them, some people will in future react to much lower doses.

14.3.1 Manipulating chemicals

Chemicals should not be touched by hand. Pupils need to be taught to remove the stopper from bottles of liquid with one hand, and keep it in their hand while pouring from the bottle. They should be told to pour on the opposite side to the label, so that it does not become damaged by corrosive chemicals. Solids should never be handled with the fingers. Instead, train pupils to use a spatula or equivalent.

14.3.2 Heating chemicals

Heating chemicals safely is a skill which pupils need to be taught, and, once taught, pupils need frequent reminders of the technique. Small quantities of a solid can be heated in test tubes; the solid should not be tightly packed and there should be an air space above the slope of the solid. Liquids present greater problems, because of the risk of 'spitting'. Boiling tubes are safer than test tubes because of their greater volume. Even so, they should not be more than about one-fifth full; anti-bumping granules may be useful. Pupils will usually remember to point test tubes away from their own face, but may have to be reminded about the need to do the same for their neighbours.

Flammable liquids should not be exposed near sources of ignition such as Bunsen burners. In particular, they should not be heated over a naked flame. If, for example, hot ethanol is required, it is most easily obtained by standing a test tube or boiling tube containing ethanol in a beaker of hot water, the hot water having been obtained directly from a tap, or from a kettle. Heating or pouring activities, especially those involving liquids, should be conducted standing up, so that the pupil or teacher can move quickly out of the way if necessary.

14.3.3 Identifying gases by inhalation

Identification of gases by their odour is an important technique and pupils should be taught how to do it safely. The gas should be contained in a test tube, not a larger vessel. The lungs should be filled with air by inhaling deeply. The test tube is held about 10-15 cm from the face, pointing away from it, and then the contents of the test tube cautiously sniffed, by using a hand to waft the vapours gently to the nose. Pupils should practise the technique with low hazard gases under supervision, and those who cannot reasonably be trusted to follow these instructions should not be permitted to progress to more hazardous substances. Teachers should discreetly check whether there are, for example, asthmatics in the class, and those affected should not smell gases such as chlorine or sulphur dioxide.

14.3.4 Spills

Spills should always be cleared up immediately. While a few may need absorption and/or chemical neutralisation using a spill kit (or disinfection or similar treatment), most minor spills can be dealt with by a damp cloth. For large spills of chemicals producing hazardous fumes, there may be a need to call the fire brigade. Pupils should be encouraged to report spills and breakages, so that they can be cleared up immediately, and not left to cause injury to the next class, or to a technician or cleaner.

14.3.5   Personal cleanliness

Pupils should always wash their hands after practical work with chemicals or with soil or material of living origin and facilities should permit this.

14.4   Glassware

14.4.1   Broken glass

Glass vessels, such as test tubes, flasks and beakers, should be checked for cracks and chips before use. Particular care should be taken when glass containers are evacuated; use round-bottomed or pear-shaped vessels, check for cracks, and protect observers by one or more safety screens. Broken glassware should be placed in a specially labelled bucket. It should be wrapped in newspaper or sealed in a non-perforatable container, such as a box or metal can, before disposal with normal waste.

14.4.2   Handling glass

When inserting corks, stoppers or bungs into test tubes or specimen tubes, pipettes into safety fillers or glass tubing or thermometers into bungs, pupils need to be shown a safe technique (see Handbook or Safeguards). Inserting glass tubing or thermometers into bungs is generally best left to technicians, who may themselves need training.

14.5   Safety equipment

14.5.1   Using safety equipment

A risk assessment is necessary to determine what type of PPE is appropriate.

14.5.2   Eye protection

Eye protection must conform to the relevant British Standard BS2092. Reputable suppliers will stock nothing else. Table 14.1 suggests appropriate eye protection for different situations. While goggles conforming to BS2092C offer better protection than safety spectacles, their elastic straps tend to perish rapidly and they mist up. These problems make it difficult for pupils to wear them, and in practice safety spectacles may be safer.

Eye protection needs careful storage to minimise the risk of scratching the lenses (see CLEAPSS Handbook). Safety spectacles and other eye protection should be cleaned regularly, and checked for scratches. Badly scratched lenses which obscure vision are one of the main reasons why pupils are sometimes reluctant to wear eye protection.

Table 14.1   Eye protection for different activities

Operation	Type of eye protection
Dispensing large volumes of concentrated acids, alkalis and other corrosive substances; opening containers which may be under pressure. Pupils with visual impairment or with other disabilities which require them to work closer to chemicals than most pupils; those with limited motor control.	Face shields meeting BS2092.
Activities using alkalis of molar or greater concentration; concentrated mineral acids and ethanoic and methanoic acids; bromine; corrosive solids; toxic chemicals.	Goggles meeting BS2092C.
Activities involving other chemicals which are classified as irritant or offer less risk to the eye.	Spectacles meeting BS2092.
Activities with other risks: glass working, breaking up rocks, stretching wires or cords; some dissection.	Any eye protection; for example, spectacles meeting BS2092.
Use of lasers.	No protection is advised for school-type lasers (which should be classed as 2 or 3a).
Use of ultra-violet radiation.	The most appropriate protection is to arrange equipment so that UV cannot reach the eye.

14.5.3   Gloves

Protective gloves are rarely likely to be necessary in Key Stages 3 and 4, but must be worn whenever the risk assessment requires them. Different types are needed for different purposes. Heat-resistant gloves may be needed in some situations. Disposable plastic gloves protect the skin from contamination by radioactive substances, while not providing protection from the radiation itself. Particularly for work in A-level chemistry, rubber or synthetic rubber gloves may be needed to prevent some chemicals being absorbed by the skin. Kitchen gloves are often satisfactory, but are not always impervious to the solvents used, and should be removed rapidly if concentrated acids or other corrosives fall on them (see Handbook).

14.5.4   Laboratory coats

Laboratory coats can stop ties or scarves from hanging loosely and getting into Bunsen burner flames or entangled with machinery. In most school science activities, however, they mainly serve to protect the person's clothing, rather than the person. Exceptions to this are in work with radioactive materials and in microbiology, where removal of the laboratory coat removes the contamination. Teachers, technicians and students on A-level or equivalent courses are likely to be handling chemicals so frequently that wearing a laboratory coat is common sense. A school may well encourage pupils in Key Stages 3 and 4 to purchase their own laboratory coats (or wear old clothing over the top of normal clothes) to prevent claims about damaged clothing, but it would be difficult for a school to provide coats in a sufficient range of sizes without causing major storage and washing problems. Coats or old clothing will only offer protection if buttoned up, and teachers should set a good example in this respect.

14.5.5   Impervious aprons

These should be worn when large quantities of corrosive liquids are being dispensed.

14.5.6   Protective footwear

This is not normally necessary in school laboratories, but neither staff nor pupils should wear open-toed sandals or similar shoes, which give no protection at all against spills or broken glass. Speed of removal is important, as ordinary shoes will admit spilt chemicals.

14.5.7   Protective headgear

This is unlikely to be necessary in school laboratories, but may be needed on some outside visits. Staff and pupils with long hair need to tie it or clip it back, so that it cannot enter a Bunsen flame or become entangled with machinery.

14.6   Control equipment

14.6.1   Safety screens

These are likely to be required by most risk assessments whenever:

- there is any risk of explosion or implosion;
- nylon filaments are stretched to breaking point;
- fast-moving objects such as airgun pellets could go in unintended directions; or
- there are other reasons generally for preventing observers from getting too close to equipment.

During demonstrations two or more safety screens are likely to be necessary, so that both pupils and teacher are protected. Some safety screens are unstable and may need clamping to the bench. Screens should be regarded as back-up equipment; full precautions should be taken to avoid explosions or other accidents. Nor are they a substitute for eye protection. See Handbook and Safeguards.

14.6.2   Fume cupboards

Fume cupboards should be used whenever the risk assessment requires it. (See Section 10)

14.7   Emergency equipment

14.7.1   Fire-fighting equipment

This should be checked regularly. Generally the whole school will have an arrangement with an outside contractor. Heads of science should report in writing any obvious deficiencies in provision. Schools are likely to be advised by a fire prevention officer on appropriate alarms, fire extinguishers and so on.

14.7.2 Eye washing facilities

These will be needed for use in emergencies, and staff will need training in how to use them. Plumbed-in eye washing stations are probably too expensive for school use. Eye-wash bottles contain only small volumes of water, and are not suitable unless the sterility of the water can be guaranteed, for example, by changing it weekly. A satisfactory alternative is to attach a short length of rubber tubing to a cold water tap, especially if it is labelled 'eye washing station'.

14.7.3 Spill kits

These should be available to deal with all likely eventualities. They should contain materials for absorbing spills and, if necessary, neutralising and emulsifying them before disposal. See Handbook and Safeguards.

14.7.4 Respirators

These are useful in school laboratories for clearing up large spills of toxic and volatile chemicals in areas where the ventilation is poor. However, as such spills are rare in school, and full-face respirators are expensive and the filters must be replaced every two years, the cost is difficult to justify. Therefore, it is more practicable to summon the fire brigade, making it clear that by their standards the spill is very small.

14.8 Investigative science

Open-ended, investigative or project work is required in many science courses. Whenever and in whatever context such open-ended activities take place, there is a need to plan carefully for the safety aspects.

In the first instance pupils should be encouraged, as part of the planning exercise, to assess the risks and devise suitable precautions. However, since teachers cannot rely on pupils having carried this out adequately, they must always check the plans, whether written or otherwise. Sometimes, the implications of the plans may not be clear, either to the pupil or to the teacher, and in any case plans may change as the investigation or project proceeds. Teachers' supervisory skills will be tested to the full when pupils in a class are all working on different investigations. It is therefore best to set investigations in the context of relatively safe activities, avoiding, for example, the use of particularly hazardous chemicals, dangerous equipment, or harmful micro- organisms.

15 USING CHEMICALS IN SCIENCE

15.1	Types of hazard

The main headings under which chemicals are classified are:

- very toxic. [insert symbols]
- toxic;
- harmful;
- corrosive;
- irritant;
- oxidising;
- explosive;
- extremely flammable;
- highly flammable;
- radioactive; or
- dangerous for the environment;

15.1.1 The CHIP Regulations

The Chemical (Hazard Information and Packaging) Regulations 1994 ('CHIP Regulations') specify the criteria for the labelling of dangerous chemicals, and also require suppliers of such chemicals to provide "Safety Data Sheets" for their products. The "Safety Data Sheet" must contain information, in a specified format, to enable the recipient to take the necessary measures relating to the protection of health and safety at work. Radioactive substances are excluded from these regulations.

The regulations specify a number of categories of danger which relate either to the physical properties of the dangerous substance, eg Oxidising, Highly Flammable or its health affects, eg Toxic, Harmful, Corrosive. For a large number of chemicals this information is contained in the Approved Supply List. However if the chemical does not appear in that list then the supplier must classify it in accordance with the procedures laid down in the regulations.

15.1.2 Exposure limits

The Control of Substances Hazardous to Health Regulations 1994 lay down Occupational Exposure Limits for Hazardous Substances which may be inhaled, eg gases or volatile substances and dust. These are published annually, by the HSE, in a Guidance Note (EH40; Occupational Exposure Limits.

15.1.3 Carcinogens

The CHIP regulations define three categories of carcinogen:

- Category 1 substances: these are known to have caused cancer in humans.

Some of these are banned by national legislation (shown as X in the following tables); others are certainly not recommended for school use (shown as N), and may well be banned in local codes of practice.

- Category 2 substances: there is a presumption that the chemicals are carcinogenic, generally based on animal studies. To some extent the risk depends upon whether they are hazardous by ingestion, inhalation or skin contact, their volatility, and how the chemicals would be used in schools (for example the likelihood of producing dust). Some have major educational benefits, and a degree of judgement has been exercised in classifying some as not recommended (N), and others as suitable under restricted conditions, such as teacher only (T), or use under close supervision by post-16 pupils [shown as (Y12+)].

- Category 3 substances: there is some evidence from animal studies that there may be cause for concern, but the evidence is such that in most cases schools could continue to use these chemicals, providing that they exercise caution. Bear in mind that quantities used in schools are very small, and any exposure is usually very brief.

15.1.4   Flammable liquids

There are two parameters which define the hazard posed by flammable liquids:

- the flash point is the lowest temperature at which a liquid gives off vapour in sufficient quantity to ignite with air when a spark or flame is applied;

- the auto-ignition point is the temperature at or above which the vapour from a liquid will ignite spontaneously in the presence of air.

These properties of liquids, together with others such as evaporation rate, will have been taken into consideration when general risk assessments were drawn up.

15.2   Practical procedures using chemicals

The following tables give guidance about the suitability of a range of practical work involving chemicals which may be used in school science. The advice needs to be interpreted with care and with due regard for the actual experience and competence of the pupils in the class.

Table 15.2 lists a number of mixtures which should not be made in schools, as they have been known to be the cause of serious accidents. The remainder of the tables seek to warn, but not prohibit.

Any procedure involving the use of unfamiliar apparatus or chemicals which the teacher has not handled before should be rehearsed before it is demonstrated or used as a class practical.

The absence of any procedure from the lists should not be taken to imply anything.

15.2.1 Disposal of chemical waste

Practical activities involving chemicals inevitably generate chemical waste. The disposal of such waste needs a risk assessment, which must be for the whole operation, including storage prior to disposal. Procedures should take account of the information contained in the Safety Data Sheet. It should be remembered that waste can often be made safe by dilution. For example, solids which cannot be put in the refuse in concentrated form can be mixed with sand; water-miscible liquids which when undiluted cannot be poured down the drain can be acceptable if diluted or poured down in batches with copious flushing. Disposal via lavatory pans is useful if large-scale flushing is necessary; their use avoids the risk of hazardous concentrations of contaminants being left in sink traps. Similarly, large quantities of chemicals whose disposal causes problems can often be divided into smaller batches for which disposal is safe.

There are seldom problems with washings from glassware: they can be washed down the drain. Large quantities (the contents of bottles, 500-2000 cm^3) pose special problems; however, if the contents are dealt with in batches, it may be safer than storing them for a contractor to remove, and is certainly cheaper.

For information about the disposal of specific chemicals and the details of methods of disposal, consult Hazcards, Handbook or Topics: only an outline of the methods of disposal of intermediate quantities can be given in the table opposite.

Table 15.1   Disposal of chemicals

Class of substance	Method of disposal
Special cases	For disposal of bromine, di(dodecanoyl)-peroxide, mercury, white phosphorus and silicon tetrachloride [*Hazcards, Handbook* or *Topics*].
Flammable liquids	Large quantities (the contents of bottles, 500–200 cm^3) by contractor, as quickly as possible. In some situations, it may be possible to burn off reasonable quantities in small batches in shallow pans in the open air. Water-miscible: intermediate quantities can be flushed down the drain. Water-immiscible: intermediate quantities can be flushed down the drain, after emulsifying with liquid detergent.
Flammable solids	Up to 20 g of metal powders such as magnesium and aluminium can be diluted with sand and put in the refuse. Red phosphorus is best burnt in small quantities in a safe place. See the references for sodium, potassium and other water-reactive solids. [*Hazcards, Handbook* and *Topics*]
Toxic chemicals	10 g or less of toxic salts may be dissolved, diluted a thousandfold and flushed to waste. Up to 10 cm^3 of organic liquids can be treated in this way, after emulsification with detergent if necessary. Toxic organic solids should be kept for disposal by a contractor.
Corrosive liquids	Dilute and neutralise before disposal.
Water-reactive corrosives, for example the chlorides of phosphorus.	These should be added cautiously to a large quantity of water before flushing to waste.
Corrosive solids	These should be dissolved cautiously in water, the solution diluted greatly and preferably neutralised before flushing to waste.
Oxidising agents	These should be dissolved in water and the solution diluted greatly before flushing to waste. Care should be taken to avoid contaminating wood, cloth etc.
General chemicals	Treat copper or zinc salts as toxic chemicals. Others can be disposed of through the refuse or flushed to waste as appropriate, with dilution. Very small quantities of chlorinated solvents can be disposed of down the drain after emulsification. Larger quantities should be allowed to evaporate away in a safe place or disposed of by a contractor.

Table 15.2   Mixtures which should NOT be made

Practical Procedure	Hazard(s)	Suitability	Guidance Comments	Reference
Aluminium (powder) + copper oxides	Explosion	N	This mixture can explode violently on heating.	
Aluminium (powder) + lead oxides	Explosion	N	This mixture can explode violently on heating.	*Safeguards*
Ammonia (solution) + iodine	Explosion	N	The mixture is explosive when dry.	
Chlorates(VII) + sulphur or phosphorus	Explosion	N	This mixture can explode violently on heating, or simply on mixing.	
Chlorine + hydrogen	Explosion	N	The explosion of this mixture is initiated by light, but hydrogen can be burnt in an atmosphere of chlorine.	
Ethanol + concentrated nitric acid	Explosion	N	This mixture may explode after an induction period.	
Gunpowder or other known explosives	Violent explosion	X	Making, or attempting to make, gunpowder and other explosive mixtures, is illegal without a licence.	
Lead nitrate + reducing agents	Explosion	N	This mixture can explode violently on heating.	
Magnesium (powder) + silver nitrate	Explosion	N	Explodes violently with traces of water.	
Potassium manganate (VII) + concentrated sulphuric acid	Explosion + corrosive chemical	N	Explodes on contact.	
Trichloromethane + propanone	Explosion	N	This mixture may explode after an induction period.	

Table 15.3   Procedures requiring special care

Practical Procedure	Hazard(s)	Suitability	Guidance Comments	Reference
Aerosol cans - use of	Propellant ignition & explosion; hazardous contents	(Y9+)	Use of laboratory aerosols (freezer spray, ninhydrin) should be carefully controlled. Do not allow personal aerosols.	
Biological use of chemicals	Various			*Handbook Safeguards*
Burning hydrogen in air	Risk of explosion	T	Ensure hydrogen is pure before attempting to ignite at the generator/ delivery tube. Safety screens essential.	*Topics*
Burning substances in air	Toxic gases	(F)	Normally a fume cupboard should be used, unless using **very** small quantities of substances known not to produce toxic products. See also Plastics testing and Calorimetry.	*Topics*
Calorimetry - fuel and food combustion	Explosion, especially with oxygen		Eye protection and safety screen required. Use air instead of oxygen if possible. When using oxygen, flush apparatus with the gas.	
Centrifugation	Risk of injury		Ensure that the load is counterbalanced. Do not open until the rotor is at rest.	
Chromatography	Solvent vapour	(F)	Carry out in a fume cupboard or in a sealed container unless a solvent of low volatility, for example, water, is used.	
	Locating agent		Many are hazardous. See Aerosol cans.	
Cleaning glassware using chromic acid	Chromic acid is corrosive	N	It is safer to use a commercial laboratory-glassware detergent.	
Cooling curves, melting point determination	Possibility of toxic vapour		Avoid substances with high melting temperature or harmful vapour. Naphthalene is not recommended except for demonstrations unless the tube is loosely stoppered with a plug of mineral wool. Otherwise use hexadecanol or octadecanoic acid.	*Topics Hazcard*
Diffusion of gases	Toxic gases, corrosive liquids	F	Particular care needed with bromine. Use fresh rubber tubing each time. Demonstration can be done in open laboratory with safety screens, but it is essential to use fume cupboard when dismantling. Sodium thiosulphate should always be available. Wear rubber (not disposable polythene) gloves. Beware of theft of ampoules.	*Hazcard*
Distillation atmospheric pressure	Nature of chemicals involved		Beware of suck-back and blocked apparatus, for example crude oil (use substitute)	*Topics*
reduced pressure	Implosion, scattering of glass		Check for cracks or scratches in glass which weaken it. Use safety screens.	

Table 15.3   Procedures requiring special care - *continued*

Practical Procedure	Hazard(s)	Suitability	Guidance / Comments	Reference
Drying gases	Blocked apparatus, hazardous substances		Avoid using concentrated acids wherever possible. Some solid drying agents can 'cake' together, blocking the apparatus, particularly if the drying tube is not freshly filled.	*Safeguards*
Dyeing	Some dyestuffs may be irritant		Avoid skin contact and inhalation of dust. Some may be sensitisers.	*Topics*
Electrolysis	Nature of chemical products		Electrolysis of some solutions may produce toxic gases (for example chlorine from concentrated aqueous sodium chloride). Safe in well-ventilated laboratory if quantities are small.	*Hazcard*
	Risk of burns with molten electrolytes	F	Molten solids (for example lead bromide) may give rise to harmful dust or fumes: a fume cupboard must be used.	
Explosion of gases			Use safety screens and eye protection for all. Risk of hearing damage - one moderate bang is enough.	*Safeguards*
in plastic bottle	Jet of flame	T	Do not handle bottle once lit.	
in metal can	Risk of burns	T		
eudiometry	Risk of violent explosion with unsuitable mixtures	T	Restrict to small quantities (up to $10\,cm^3$) of hydrogen/oxygen, hydrogen/air or alkane/air mixtures.	
		N	Mixtures of hydrocarbons and oxygen should not be used. Ethyne (acetylene) is particularly dangerous.	
Gas cylinders	Danger of falling cylinder causing injury and/or release of gases at very high pressure Chemical nature of the gas	F	Cylinders should be properly maintained and clamped suitably both in stores and when in use. They should be moved on a trolley. Set gas flow before connecting to glass apparatus. Certain gases such as oxygen may cause slow deterioration of components.	*Topics Safeguards*
Gas syringes	Nature of the chemicals		Ensure that the piston moves freely, attach a cord to ensure that it cannot fall out of the barrel. Take care that the capacity of the system is sufficient to contain the maximum gas volume that will arise - particularly if heating. Use safety screens if there is a risk of explosion.	

Table 15.3   Procedures requiring special care - *continued*

Practical Procedure	Hazard(s)	Suitability	Guidance Comments	Reference
Heating gases	Risk of explosion if apparatus is sealed. Escape of gas into room	(F)	Use of fume cupboard may be required (for example, with nitrogen dioxide, hydrogen halides). Be aware of the nature of a particular gas.	
Heating liquids	Hot liquids spurting (bumping)		Test-tubes should never be more than 1/5 full; it is preferable to use a wide tube (boiling tube). Always use a small flame, constantly shake the tube, and point it away from other people. Consider using anti-bumping granules. Take particular care with sodium hydroxide solution, or mixtures containing it.	*Topics* *Hazcard* *Safeguards*
Heating liquids (flammable) for example ethanol	Risk of fire/burns		Use bath of hot water (or oil) for heating or a purposely designed electric heater. All flames must be extinguished on the bench where the heating takes place. Kettles are are useful source of hot water.	*Safeguards* *Hazcards*
Heating solids	Blockage leading to explosion or forcible ejection of solid by formation of gas		The tube should be only partially filled, in a shallow layer along its length to prevent blockage.	*Safeguards*
	Toxic gases	(F)	For example, nitrogen dioxide from nitrates, sulphur oxides from sulphates.	*Hazcard*
	Thermal decomposition may be violent,	(F)	For example, ammonium dichromate. See section on chemicals.	
melting points	or produce harmful vapours	(F)	For example, depolymerisation of some plastics; see Plastics testing. See Cooling curves.	
Named reagents for example, Brady's reagent Fehling's solution Devarda's alloy	Hazard cannot be assessed unless chemical composition is known		Some named reagents have corrosive or toxic components, for example, Fehling's solution contains 4 M sodium hydroxide solution. Brady's reagent may contain either concentrated sulphuric or phosphoric acids.	*Handbook* or the supplier's catalogue
Oxidation of ethanol with acidified dichromate(VI)	Risk of reaction mixture spurting from the apparatus	(Y12+)	Mix the reagents very thoroughly keeping the mixture cool; finally raise the temperature slowly on a water bath.	
Pipettes, use of	Risk of chemicals in the mouth	N	Never pipette by mouth: always use a safety device.	
Plastics testing	Risk of toxic and flammable vapours	(F) (Y9+)	The thermal decomposition of pvc leads to the formation of chloroethene (vinyl chloride) which is an established carcinogen. Burning tests for the identification of plastics that could include pvc should be conducted on a very small scale in a fume cupboard.	

Practical Procedure	Hazard(s)	Suitability	Guidance Comments	Reference
Polymerisations addition with catalyst condensation	Risk of violent reaction/possible harmful nature of the chemicals		Di(benzenecarbonyl) peroxide (benzoyl peroxide), some times used as a catalyst in several organic polymerisations, has been the cause of explosions.	*Safeguards*
Preparation of gases	Nature of chemicals	(F)	Check if fume cupboard required. Check for blockages in the apparatus. Be prepared to avoid suckback of water or other liquids e.g. in a wash bottle by use of a trap.	
chlorine	Toxicity Risk of explosion with wrong reagent	F	A fume cupboard should be used. Several accidents have been reported in which concentrated sulphuric acid was used inadvertently in place of hydrochloric acid, leading to an explosion.	*Safeguards*
hydrogen	Risk of explosion if heating or burning		Hydrogen/air mixtures are explosive over the range of 4 to 75% hydrogen. Always ensure that the hydrogen is pure before attempting to ignite at the generator or delivery tube.	*Safeguards*
oxygen	Risk of explosion if using potassium chlorate(V) and manganese(IV) oxide	N Y7+	This mixture has given rise to violent reactions when traces of carbon or organic matter were present. It is safer to use '20 volume' hydrogen peroxide solution with manganese (IV) oxide catalyst	*Safeguards* *Topics* *Hazard*
other gases	Various			
Thermite reactions	Risk of burns	T	These reactions can be very vigorous and shower sparks over several metres.	*Topics* *Safeguards*

Table 15.4   Specific chemicals

Substance	Hazard	Suitability		Guidance Comments
Acetonitrile	Highly flammable and toxic	N	F	Benzonitrile offers a much safer alternative for reactions of this class of compounds.
Aluminium (powder – pyrophoric)	Spontaneously flammable	T		Should not be mixed with lead oxide or copper oxides. For use in the 'thermite' reaction consult section 15.2.3.
(powder – fine)	Flammable	Y7+		
Aluminium bromide (anhydrous)	Corrosive	Y12+	(F)	Use a fume cupboard for investigations of reaction with water as large volumes of corrosive gas could be produced.
Aluminium chloride (anhydrous)	Corrosive	Y12+	(F)	Violent reaction with water. There can be a long induction period with the reaction especially if the sample is old.
4-aminobenzene sulphonic acid (sulphanilic acid)	Harmful	Y12+		As precursor for the preparation of methyl orange only, not as part of the spot test for nitrites; one of the other reagents used in this test is banned in schools.
4-aminobiphenyl	Toxic	X		Banned in schools.
Aminobutanes and aliphatic amino compounds	Highly flammable and irritant	Y12+	F	See also methylamine.
Ammonia gas	Toxic	Y7+	(F)	See section 14.3.4 for details of smelling gases safely.
Ammonia solution ≥35% ammonia (0.880)(≥21 M)	Corrosive	Y12+	F	Beware of evolution of ammonia gas.
≥10% <35% (≥6 M <21 M)	Corrosive	(Y9+)	F	
≥5% <10% (≥3 M <6 M)	Irritant	Y7+	(F)	
Ammoniacal silver nitrate solution (Tollen's reagent)	Explosive	(Y9+)		Use a clean test tube which should be heated on a water bath only. Do not allow to boil dry as the solid product is explosive. Discard any residues into large volumes of water and wash the tube out with dilute nitric acid. Make only as needed, do not store the solution and do not add excess solution or product of reaction to silver residue bottles.
Ammonium chlorate(VII) (perchlorate)	Oxidising	N		Not recommended for use in schools.
Ammonium chloride	Harmful	Y7+		Fumes are irritating to the eyes.

Table 15.4   Specific chemicals – *continued*

Substance	Hazard	Suitability		Guidance Comments
Ammonium dichromate(VI)	Explosive and irritant	N	(F)	Never heat with reducing agents such as magnesium or other metals.
		Y12+		Use a fume cupboard for the 'volcano experiment'. If the solid is heated in a test tube the reaction should be carried out in a fume cupboard with a very loose-fitting glass wool plug in the end of the tube to prevent the escape of chromium(III) oxide dust. 1g of solid is the maximum that should be heated in a tube.
Ammonium ethanedioate (oxalate)	Harmful	Y12+		
Ammonium metavanadate	Toxic	Y12+		
Ammonium molybdate	Harmful	(Y9+)		Use only as phosphate-testing reagent. A solution of the reagent may contain concentrated nitric acid and this should be considered as the major hazard.
Ammonium nickel sulphate	Harmful	(Y9+)		See nickel sulphate.
Ammonium nitrate	Oxidising	N		The solid forms explosive mixtures with reducing agents and potassium manganate(VII).
		Y12+		The solid should not be heated as a preparation for dinitrogen monoxide (nitrous oxide). For details of this preparation see the section on dinitrogen monoxide.
Ammoniun nitrite	Explosive	T		This material can only be made *in situ*. The solution should never be heated to dryness.
Ammonium peroxodisulphate (persulphate)	Oxidising and harmful	Y12+		Short shelf-life; builds up pressure owing to decomposition.
Ammonium polysulphide solution (ammonium sulphide solution)	Corrosive	Y12+	F	Reacts with acid to give toxic hydrogen sulphide.
Ammonium thiocyanate	Harmful	N		Do react with concentrated sulphuric acid or boil with dilute acids; a very noxious gas is evolved. Do not heat the solid.
		Y9+		Use only in solution as a test for iron(III). Do not heat the result of the test to dryness.
Anthracene	Irritant	N		Keep exhibition sample only.
Antimony	Harmful	N		Keep exhibition sample only.
Antimony compounds	Harmful	(Y12+)		The oxide is a suspected carcinogen.

Table 15.4   Specific chemicals – *continued*

Substance	Hazard	Suitability		Guidance Comments
Aqua Regia	Corrosive	T	F	Make and use at once in a fume cupboard. Do not store.
Arsenic	Irritant	N		Keep as exhibition sample only.
Arsenic compounds	Very toxic	N		Not recommended for use in schools.
Asbestos – all forms	Toxic	N		Asbestos should not be used in any practical activities.
Azo dyes	Some of these are thought to be carcinogenic	N (Y12+)		Do not try to isolate the solid dyes. The azo dye methyl orange is water soluble and is much safer to prepare. See *Topics*.
Barium	Highly flammable	T		Handle with care; the metal is very difficult to cut.
Barium salts (general) as solids as solution (0.05 M or more)	Harmful Harmful	(Y9+) (Y7+)		
Barium chromate(VI)	Toxic	N		Not recommended for school use as a solid. If a precipitate is formed in a reaction it should not be isolated and dried.
Barium hydroxide	Harmful	Y12+		
Barium nitrate	Oxidising and harmful	Y12+		The thermal stability of barium nitrate is such that it does not easily decompose to give nitrogen dioxide and oxygen and cannot be used in place of lead nitrate as a source of a mixture of gases for separation of the nitrogen dioxide.
Barium peroxide	Oxidising and harmful	Y12+		Vigorous reaction with water to give corrosive solution. The reaction with ethanoic acid is explosive.
Barium sulphate	Minimal hazard	Y9+		The dust of this material may present a hazard.
Benedict's reagent	Minimal hazard	(Y7+)		The mixture should be heated on a water bath as test for reducing sugars. This is a much safer alternative to Fehling's solution.
Benzene	Highly flammable and toxic	X		Banned in schools. Alternatives are available; see *Hazcard*.
Benzenecarbaldehyde (benzaldehyde)	Harmful	Y12+	(F)	
Benzene carbonyl chloride (benzoyl chloride)	Corrosive	Y12+	F	The substance is lachrymatory.

Table 15.4   Specific chemicals – *continued*

Substance	Hazard	Suitability		Guidance Comments
Benzene-1,2-dicarboxylic anhydride (phthalic anhydride)	Irritant	Y12+		The vapour above the molten material presents the hazard.
Benzene-1,2-diol (catechol)	Harmful	Y12+		This material should not be reacted with concentrated nitric acid.
Benzene-1,3-diol (resorcinol)	Harmful	(Y9+)		This material should not be reacted with concentrated nitric acid. This material is generally used as a test reagent in ethanol, ethanoic acid or propanone. Treat as though the solvent were the hazard.
Benzene-1,4-diol (quinol)	Harmful	(Y9+)		This material should not be reacted with concentrated nitric acid.
Benzene sulphonic acid	Corrosive	Y12+		
Benzene-1,2,3-triol (pyrogallol)	Harmful	(Y9+)		For absorption of oxygen, prepare solution as it is needed. A solution made up with sodium hydrogencarbonate is much safer than with sodium hydroxide.
Benzene-1,3,5-triol (phloroglucinol)	Harmful	Y12+		
Benzoin	Minimal hazard	Y12+		
Benzonitrile	Harmful	Y12+	F	This compound provides a safer alternative to acetonitrile if the reactions of nitriles are investigated.
Beryllium	Very toxic	N		Keep as exhibition sample only in safe place.
Beryllium compounds	Very toxic	N		Not recommended for use in schools.
Bismuth	Minimal hazard	Y12+		
Bismuth chloride	Irritant	Y9+		
Bismuth nitrate	Oxidising and irritant	Y12+		
Biuret solution	Corrosive and irritant	(Y9+)		The main problem associated with the use of this reagent is that it contains dilute sodium hydroxide solution and should be treated as such.
Boric acid	Minimal hazard	Y7+		

Table 15.4   Specific chemicals – *continued*

Substance	Hazard	Suitability		Guidance Comments
Bromine liquid and non-aqueous solutions ≥7%	Corrosive and very toxic	(Y12+)	F	Use in a fume cupboard under very strict supervision. Suitable gloves and eye protection should be worn. Always have a solution of sodium thiosulphate available close by when this material is used. Do not react liquid bromine with concentrated ammonia solution or reactive metals. The reaction with sodium hydroxide is very exothermic.
Bromine water saturated	Toxic and corrosive	Y9+	(F)	The reaction with ethene is acceptable as little or no 1,2-dibromoethane is formed.
0.006 M to saturated	Harmful	(Y7+)		
Bromobenzene	Flammable and irritant	Y12+		This is not a suitable compound for sodium fusion.
1-bromobutane	Highly flammable and harmful	Y12+	(F)	Do not distil to completion. Steam distillation does not present any problems.
2-bromobutane	Highly flammable and harmful	Y12+	(F)	Do not distil to completion.
Bromoethane	Harmful	Y12+	(F)	Do not distil to completion.
Bromomethane	Toxic	N	F	Not recommended for use in schools.
2-bromo-2-methylpropane	Highly flammable and harmful	Y12+	(F)	Do not distil to completion.
Bromopropanes	Harmful	(Y12+)	F	Do not distil to completion.
3-bromoprop-1-ene (allyl bromide)	Highly flammable and very toxic	T	F	Use only in a fume cupboard.
Butanal	Highly flammable	Y9+	(F)	
Butane gas in cylinder	Highly flammable	N		The use of large cylinders of butane is not recommended in schools other than as fuel for a temporary heat source or piped gas supply.
		(Y9+)		However, small cylinders to which a regulator and tube can be fixed or gas cigarette lighters are useful sources of the gas for experimentation purposes, provided they are used only under close supervision.
Butanoic acid	Corrosive	Y12+		This material has a very unpleasant smell. If it is used to produce esters, it is advisable to make the reaction mixture in a fume cupboard.
Butanols	Highly flammable and harmful	(Y7+)	(F)	Heat only on a water bath.

Table 15.4   Specific chemicals – *continued*

Substance	Hazard	Suitability		Guidance Comments
Butanone	High flammable and irritant	Y9+	(F)	Heat only on a water bath.
cis-butene-1,4 dioic acid and its anhydride (maleic acid and anhydride)	Harmful	Y12+		The anhydride can cause sensitisation by inhalation.
trans-butene-1,4-dioic acid (fumaric acid)	Irritant	Y12+		
Cadmium	Toxic	T		Keep only as exhibition sample. Nicad batteries should not be opened.
Cadmium salts (general)	Toxic	(Y12+)		Samples of cadmium salts are generally left-overs from classical analysis. There is no point in buying new supplies of these toxic materials. The electrolysis of molten cadmium salts is not recommended. If cadmium salts are used, care needs to be taken when they are disposed of.
Caffeine	Toxic	Y12+	(F)	Instructions for the extraction of caffeine from tea or coffee generally suggest using trichloromethane. Dichloromethane provides a safer alternative. The extraction should be carried out in a fume cupboard unless very small volumes are involved.
Calcium	Highly flammable	(Y9+)		The metals should not be reacted with sulphur or the hydroxides or carbonates of the alkali metals.
Calcium chlorate(I) (bleaching powder)	Oxidising and corrosive	(Y9+)		Useful as a safer source of chlorine. Old stock can become dangerous if traces of metals or rust are present.
Calcium chloride (hydrated and anhydrous)	Irritant	Y7+		
Calcium dicarbide (calcium carbide)	Flammable	(Y12+)	(F)	If the solid is to be used to generate ethyne, large lumps and not powder, should be used and the gas produced treated with extreme care as the resulting mixture of ethyne and air is very explosive. The reaction between the dicarbide and solutions of silver or copper salts produces explosive acetylides.
Calcium hydroxide (solid)	Irritant	(Y7+)		
Calcium oxide	Corrosive	(Y7+)		
Calcium phosphide	Highly flammable and very toxic	N		This material is not recommended but if kept, it should be stored in a dry place with a rubber stopper.

Table 15.4   Specific chemicals – *continued*

Substance	Hazard	Suitability		Guidance Comments
Calcium salts (general)	Minimal hazard	Y7+		Unless the anion dictates otherwise, most calcium compounds present minimal hazard.
Camphor	Harmful	(Y9+)		
Carbon dioxide Solid		Y12+		To avoid chill-burns, always use tongs and thick gloves when handling the solid.
Cylinder		Y12+		
Carbon monoxide Cylinder	High flammable and toxic	N	F	The gas should not be used for reduction of heated metal oxides unless the same precautions that would be applied to the use of hydrogen are in force.
Prepared in reaction		Y12+		
Carbon disulphide	Highly flammable and toxic	N		Not recommended for use in schools because of possible risk of impaired fertility and harm to the unborn child. Also, the vapour can spread some distance and no heat sources of any type are safe in the same laboratory. Use dimethylbenzene for preparation of the allotropes of sulphur. Ethyl 3-phenyl-propenoate (ethyl cinnamate) can be used in 'hollow prism' experiments. The prism needs thorough cleaning with propanone after use.
Cement and mortar	Corrosive (irritant dust)	Y7+		Treat as an alkaline substance. The dust probably presents the main hazard.
Chlorates(I), (V) and (VII) (hypochlorites, chlorates and perchlorates)				See entries under calcium, potassium and/or sodium salt and chloric(VII) acid.
Chloric(VII) acid (perchloric acid)	Oxidising and corrosive	N		

Table 15.4   Specific chemicals – *continued*

Substance	Hazard	Suitability		Guidance Comments

Chlorine
   Cylinder
 	
   Preparation for use in 'large-scale' demonstration
 	
   Preparation for use in 'small-scale' demonstration
 	
   Preparation in test tubes
	Toxic	N
		
		T
		
		(Y12+)
		
		(Y9+)
			
			F
				May trigger an asthmatic attack.
				Not suitable for use in schools.
				Chlorine should not be reacted with fine metal powders such as aluminium, hydrocarbon gases or with ammonia or its compounds. Mixtures of hydrogen and chlorine gases are explosive over a range of compositions.
				There have been many accidents involving the preparation of chlorine where concentrated sulphuric acid (used to dry the gas) and concentrated hydrochloric acid (used to prepare the gas) have been confused. Safer methods are available such as the action of acid on chlorate(I) solution.
				Students should be taught how to smell gases safely before they are asked to smell chlorine.

Chlorobenzene	Flammable and harmful	Y12+	F	This material is not suitable for sodium fusion.
Chlorobutanes (butyl chlorides)	Highly flammable	Y12+	F	Do not distil to completion.
Chloroethane (ethyl chloride)	Highly flammable	Y12+	F	It boils at 12°C.
Chloroethene (vinyl chloride monomer)	Highly flammable and toxic	N		Not suitable for use in schools: it is an established carcinogen. See also section 15.2.3, 'Plastics testing'.
Chloroethanoic acids	Toxic	(Y12+)		These materials should not be heated to dryness as they can decompose to give toxic phosgene.
Chloromethane	Highly flammable and harmful	N		This, and other low molecular mass haloalkanes, are not recommended for use in schools.
(Chloromethyl)benzene (benzyl chloride)	Irritant	Y12+		This material has been safely used for sodium fusions in a fume cupboard.
Chloroplatinic acid	Corrosive	T		The dust may trigger an asthmatic attack. It may cause sensitisation by skin contact.
Chloropropanes (propyl chlorides)	Highly flammable and harmful	Y12+	(F)	
Chlorosulphonic acid	Corrosive	N		This substance has a violent reaction with water and is therefore not recommended for use in schools.

Table 15.4   Specific chemicals – *continued*

Substance	Hazard	Suitability	Guidance Comments
Chromates(VI) and dichromates(VI)			
Water-insoluble solids	Irritant	N	Chromate precipitates should not be isolated and dried in experiments. In the oxidation of alcohols, ensure that all the solid has dissolved and that the contents of the flask are thoroughly mixed before refluxing. See also individual chromates.
Water-soluble solids	Irritant	Y12+	
Solutions ≥0.025 M	Irritant	Y9+	
'Chromic acid' cleaning mixture	Corrosive	N	This mixture of concentrated sulphuric acid and potassium dichromate used for cleaning glassware reacts violently with many organic compounds and residues. Under no circumstances should the material be stored. Commercial detergents are much safer.
Chromium(III) compounds	(Irritant) (Harmful)	Y9+	No hazard categorisation under CHIP Regulations. Some suppliers do issue warnings which may not be consistent.
Chromium(VI) oxide (chromium trioxide)	Toxic, oxidising and corrosive	N	May cause cancer by inhalation of dust and may cause sensitisation by skin contact. This material shows a violent reaction with materials such as alcohols, ketones, ethanoic acid and glycerol as well as assisting combustion of wood and fabric. If possible a safer alternative should be found.
		(Y12+)	If chromium(VI) oxide is essential as part of a 'preparation' for biological use, gloves and eye protection should be worn and the preparation should be made up in a fume cupboard.
Chromium(VI) dichloride dioxide (chromyl chloride)	Oxidising and corrosive	(Y12+) F	Preparation of this material should be carried out only in small quantities. Any excess material should be reacted with dilute alkali solution with care.
Coal	Minimal hazard	(Y9+) (F)	Destructive distillation produces toxic gases, some of which may be burnt off or absorbed in water. Good ventilation is essential.
		T	Care is needed when cleaning glassware containing tarry residues.
Cobalt chloride			This material may cause skin sensitisation. There is inconsistent information from suppliers about the hazard category.
Solid	Harmful	(Y7+)	
Solution ≥1 M	Harmful	(Y7+)	
Colchicine	Very toxic	N	Not suitable for use in schools. 1.4-dichlorobenzene is a safer alternative (see *Hazcards*).

Table 15.4   Specific chemicals – *continued*

Substance	Hazard	Suitability		Guidance Comments
Copper (II) salts (general) e.g. sulphate				There is a danger that the toxicity of these materials can be underestimated because of their familiarity.
Solids	Harmful	Y7+		
Solutions ≥1.4 M	Harmful	Y7+		
Copper(II) chloride				
Solid	Toxic	Y12+		
Solution ≥1.4 M	Toxic	(Y9+)		
≥0.15 M <1.4 M	Harmful	Y9+		
Copper(I) oxide	Harmful	Y7+		
Copper(II) oxide	Harmful	N		This material is not safe in the 'thermit'.
		T		Great caution is needed when reduced with magnesium or hydrogen.
		(Y7+)		For use as a base and reactions with zinc or carbon.
Crude oil	Highly flammable and toxic	N		Crude oil should not be used in school as it is carcinogenic. In the past it has been possible to purchase material that has been stripped of volatile materials and water. This is referred to as stabilised crude. (see section 15.2.3 for safe ways to distil such mixtures.) Untreated materials contain low boiling point liquids and dissolved gases which boil out with no warning and present a fire hazard.
(Simulated 'crude oil' mixture)	Highly flammable	(Y7+)	(F)	There is an excellent recipe in *Hazcards* which is safe to distil in small quantities by younger students under supervision or for demonstrations. Care is needed when cleaning the tarry residues.
Cyanides				Not recommended for use in schools. Solutions of cyanides are sometimes used in advanced biological experiments.
Solid	Toxic	N		
Solution		T		It is safest if ready-prepared solutions are bought in as and when they are needed and used only by staff.
Cyclohexane	Highly flammable	(Y9+)	(F)	Small-scale experiments on the reactions of saturated hydrocarbons can be carried out in a well-ventilated laboratory.
Cyclohexanol	Harmful	Y12+	(F)	
Cyclohexanone	Flammable and harmful	Y12+	(F)	
Cyclohexene	Highly flammable and irritant	Y9+	(F)	This material can be used to compare saturated and unsaturated compounds but care needs to be exercised in the disposal of the liquid. The material may form unstable peroxides on storing over a long period.
DCPIP	Minimal hazard	Y7+		

Table 15.4   Specific chemicals – *continued*

Substance	Hazard	Suitability		Guidance Comments
DDT	Toxic	N		
Decanedioyl dichloride (sebacoyl chloride)	Corrosive	Y12+	F	If Nylon is to be produced, the solvent cyclohexane is safer than the halogenated hydrocarbons. It forms the upper layer if the diaminohexane is used in water.
Use in Nylon preparation		(Y9+)	(F)	Very small scale if no fume cupboard used.
Devarda's alloy	Minimal hazard	Y9+		Strongly exothermic reaction with sodium hydroxide solution.
1,2-diaminoethane	Flammable and corrosive	Y12+		Sensitiser.
Di(benzenecarbonyl)peroxide (benzoyl peroxide)	Explosive and irritant	N		Di(dodecanoyl)peroxide provides a much safer alternative.
1,2-dibromoethane (ethylene dibromide)	Toxic	N		Not recommended in schools as it is carcinogenic. However, the reaction between bromine water and ethene does not present a risk as the amount of dibromide formed is very small. No attempt should be made to isolate the dibromide.
1,2-dibromopropane (propylene dibromide)	Harmful	Y12+	F	
Dichlorobenzenes	Harmful	Y12+		1,2-dichlorobenzene has been used to raise leatherjackets from the soil; ethanal will do in its place and 1,4-dichlorobenzene is an alternative for the pretreatment of roots to arrest metaphase. (See *Hazcards*.)
Dichlorobiphenyl-4,4'-diamines (chlorobenzidines)	Toxic	X		Banned in schools.
1,2-dichloroethane (ethylene dichloride)	Highly flammable and toxic	N		Not recommended in schools, as it is carcinogenic.
Dichloroethanoic acid (dichloracetic acid)	Corrosive	Y12+		
Dichloromethane	Harmful	(Y9+)	(F)	
Di(dodecanoyl)peroxide (lauroyl peroxide)	Oxidising and irritant	(Y9+)		The reaction with methyl methacrylate to produce Perspex is very exothermic if the concentration of the peroxide is >1%. The solid should not be allowed to dry out or be heated on its own.
Diethylamine			F	
Liquid	Highly flammable and irritant	(Y12+)		
Solution		(Y9+)		

Table 15.4   Specific chemicals – *continued*

Substance	Hazard	Suitability		Guidance Comments
Diethyl sulphate	Toxic	N		Not suitable for use in schools.
Diiodine hexachloride (iodine trichloride)	Corrosive and toxic	(Y12+)	F	
3,3'-dimethylbiphenyl-4,4'-diamine (*o*-tolidine) and its salts	Toxic	X		Banned in schools.
Dimethyl formamide	Harmful	Y12+	F	
Dimethyl sulphate	Very toxic	N		Not suitable for use in schools.
Dinitrobenzene	Toxic	(Y12+)		
3,5-dinitrobenzoic acid	Harmful	Y12+		
4,4'-dinitrobiphenyl	Toxic	X		Banned in schools.
2,4-dinitrobromobenzene	Toxic	N		Not suitable for use in schools. Sensitiser.
2,4-dinitrochlorobenzene	Toxic	N		Not suitable for use in schools. Sensitiser.
2,4-dinitrofluorobenzene	Toxic	N		Not suitable for use in schools. Sensitiser.

Dinitrogen oxide (nitrous oxide)
   Cylinder
   Prepared for reactions
	Toxic	
		N
		Y12+
				This gas should not be prepared by the thermal decomposition of ammonium nitrate. Either ammonium sulphate/ potassium nitrate or hydroxylamine hydrochloride/iron(III) ammonium sulphate mixtures are better alternatives. The gas is explosive if mixed with ammonia, carbon monoxide, hydrogen or hydrogen sulphide.

Dinitromethylbenzene (dinitrotoluene)	Toxic	Y12+		
Dinitrophenols	Toxic	(Y12+)		It is possible that these compounds may be prepared by side reactions during the nitration of phenol. They should not be isolated and care needs to be taken when apparatus is being washed. The hazards associated with samples of the mixture produced for use in chromatographic analysis can be reduced if gloves are worn, the reaction carried out on a micro scale and the tubes wrapped and discarded after use.
2,4-dinitrophenylhydrazine	Toxic and explosive	Y12+		The solid should be kept damp at all times. Gloves need to be worn when using this solid.

Table 15.4   Specific chemicals – *continued*

Substance	Hazard	Suitability		Guidance Comments
1,4-dioxane	Highly flammable and irritant	(Y12+)		May form peroxides on standing. Short shelf-life.
Dipentene (limonene)	Irritant	Y9+		Cyclohexane or dichloromethane may be used in place of 1,1,1-trichloroethane to extract limonene from the aqueous solution after the steam distillation of orange peel.
Diphenylamine Solid Solution	Toxic	T (Y12+)		The best advice is to buy purified material as impure samples may contain carcinogens. The toxic effects of the material are cumulative. If the material is used in concentrated sulphuric acid, care needs to be exercised in making the solution.
Disulphur dichloride	Corrosive	Y12+	F	Reacts violently with water producing a range of sulphur compounds. The use of a fume cupboard is essential.
Dodecylbenzene sulphonic acid (Nansa acid)	Corrosive	Y9+		The main hazard is its viscosity which can make transfer of the liquid very difficult.
Dye stuffs Solid Solutions	Irritants	T Y9+	F	Some sensitisers, for example, Procion dyes. The dust produced by some of these materials may cause problems. Students should not be allowed to prepare their own solutions for dyeing experiments. Staff preparing the solutions should do so in a working fume cupboard.
Ethanal (acetaldehyde)	Extremely flammable and harmful	Y12	(F)	Use a water bath for heating. Small quantities of acid or alkali can cause the material to polymerise violently. It is best to keep small quantities for class use in separate, appropriately-labelled bottles.
Ethanal tetramer (metaldehyde)	Flammable	(Y9+)		This is a safe alternative to alcohol for steam engines. It produces a toxic vapour if heated but not when burnt.
Ethanal trimer (paraldehyde)	Highly flammable	Y12+	F	
Ethanamide (acetamide)	Harmful	Y12+		
Ethane cylinder	Highly flammable	N		
Ethanedioic acid and salts (oxalic acid) Solid Solution ≥0.3 M	 Harmful Harmful	 Y12+ Y9+		Action of concentrated sulphuric acid on these substances produces carbon monoxide.
Ethanediol (ethylene glycol)	Harmful	Y12+		

Table 15.4   Specific chemicals – *continued*

Substance	Hazard	Suitability		Guidance Comments
Ethanoic acid (acetic acid)				The liquid acid reacts violently with chromium(VI) oxide, manganate(VII), nitric acid and peroxides.
Glacial	Corrosive	Y9+	(F)	A fume cupboard should be used if large volumes are involved, for example in the preparation of dilute solutions. Gloves and eye protection are vital when handling the liquid.
≥M	Corrosive	Y9+		
≥1.5 M <4 M	Irritant	(Y7+)		
Ethanoic anhydride (acetic anhydride)	Flammable and corrosive	Y12+	(F)	Use a fume cupboard if large volumes are involved. The reaction with water can be hazardous. Heat builds up at the interface between the two liquids. This reaction is catalysed by acids such as ethanoic acid. The reaction with ethanol can be very vigorous in the presence of acids. The liquid reacts violently with boric acid, chromium(VI), oxide, manganate(VII), nitric acid and peroxides.
Ethanol and Industrial Methylated Spirits	Highly flammable	(Y7+)	(F)	Use a fume cupboard if large volumes are involved. All experiments involving heating, for example, the extraction of chlorophyll, should be carried out on a water bath, in a beaker of hot water (heated in a kettle) or using an electric heater with a sealed element. Several serious accidents have occurred when this material has been used as the fuel for steam engines or in experiments where heats of combustion are being measured. The accidents tend to occur when material is being added to a burner that had been thought to be extinguished or liquid is being decanted too close to a naked flame. For steam engines, solid fuels should be used to replace the alcohol. For combustion reactions, careful supervision is essential at all times. The oxidation of ethanol with acidified dichromate(VI) is a very exothermic reaction, (see 15.2.3). Potassium reacts violently with ethanol.
Ethanoyl chloride (acetyl chloride)	Highly flammable and corrosive	Y12+	F	Violent reaction with water. The reaction with ethanol has a short induction period and can produce enough heat to eject the contents of the tube. The reactions with concentrated ammonia and with phenylamine are vigorous and produce copious amounts of white smoke. All reactions involving the acid chloride should be carried out in a fume cupboard and the use of a safety screen is a wise precaution.
Ethanoyl hydroxybenzoic acid	Harmful	(Y9+)		
Ethene cylinder	Highly flammable	N		Not suitable for use in schools.

Table 15.4   Specific chemicals – *continued*

Substance	Hazard	Suitability		Guidance Comments
Ethers (general)	Extremely flammable	(Y12+)	F	Use under strict supervision with all students.
Ethidium bromide	Very toxic	N		Not suitable for use in schools.
Ethoxyethane (diethyl ether)	Extremely flammable	(Y12+)	F	Use under strict supervision with all students. If ether is used as a solvent or in solvent extraction experiments, all sources of ignition should be extinguished. Any pressure build up in the separating funnel should be vented. Evaporation of excess ether is best carried out under reduced pressure using a side-arm flask and a bath of hot water. Sodium used to dry ether should be disposed of with care. Ethers need to be tested for peroxide build up in old stock at least once a year. This effect can be minimised by storing the ether in a dark bottle with the minimum of air space. Ethers can be tested for peroxides by the addition of a small volume of acidified potassium iodide solution. The production of iodine suggests a peroxide build up. Peroxides can be removed in several ways but these also remove added inhibitors. Methods include shaking with about half the volume of 1 M iron(II) sulphate solution. If ethers are treated in this way, they should be tested again for peroxides and used at once, not stored.
Ethyl benzoate	Minimal hazard	Y9+	(F)	
Ethyl ethanoate (acetate)	Highly flammable	Y9+	(F)	Use a fume cupboard if large volumes are involved. This material is an alternative to propanone in experiments involving the measurement of the strength of hydrogen bonds.
Ethyl methanoate (formate)	Highly flammable	Y9+	(F)	Use a fume cupboard if large volumes are involved.
Ethyl propanoate	Highly flammable	Y9+	(F)	Use a fume cupboard if large volumes are involved.
Ethyne cylinder	Highly flammable	N		Not suitable for use in school laboratories.
Fehling's Solution No 1	Harmful	Y7+		Treat as copper sulphate solution.
Fehling's Solution No 2	Corrosive	(Y9+)		Younger students need very careful supervision with this solution which contains sodium hydroxide at greater than normal bench concentration. When reducing sugars are being tested, the mixture should be heated on a water bath. Benedict's solution is a much safer alternative.

Table 15.4　　Specific chemicals – *continued*

Substance	Hazard	Suitability		Guidance Comments
Fixatives 　Solid	Some may be harmful. For full details see suppliers' data sheet.	T	(F)	These should be made up by staff. If large amounts of powder are involved, the weighing is best carried out in a fume cupboard.
Solution		Y7+		If they are used in alcohol solution they should be treated as alcohol from a safety point of view. Care needs to be exercised with old samples if their exact composition is not known. Treat as a hazard. See *Handbook*.
Fluorides Solid Solution	Toxic Toxic	T Y12+		Use only by staff.
Fluorine	Toxic	N		Not suitable for use in schools.
Fuchsin 　Solid 　Solution in ethanol	 Harmful Flammable	 T Y12+		
Germanium tetrachloride	Corrosive	Y12+	F	Very reactive with water.
Heptane	Highly flammable	Y7+	(F)	
Hexacyanoferrates(II) (ferrocyanides) 　Solid 　Solution	 Minimal hazard Minimal hazard	 Y12+ Y9+		Forms explosive mixtures with nitrites and gives hydrogen cyanide with concentrated acids. Do not heat the solid.
Hexacyanoferrates(III) (ferricyanides) 　Solid 　Solution	 Minimal hazard Minimal hazard	 Y12+ Y9+		May explode with ammonia, forms explosive mixtures with nitrites and hydrogen cyanide with concentrated acids. Do not heat the solid.
Hexamine	Flammable and irritant	Y9+		The preferred alternative to ethanol as a fuel for model steam engines: needs special burner.
Hexane	Harmful and highly flammable	Y9+	F	This material is present in low boiling point petroleum ethers. Use alternatives where possible.
Hexane-1,6-diamine (hexamethylenediamine) 　Solid 　Solution	Corrosive and harmful 	 Y12+ (Y9+)		
Hexanedioic acid (adipic acid)	Irritant	Y9+		
Hexanedioyl dichloride (adipoyl chloride). Use in preparation of Nylon	Corrosive	Y12+ (Y9+)	F (F)	The preferred solvent for the preparation of Nylon is cyclohexane.
Hexan-1-ol	Harmful	Y9+	(F)	

Table 15.4 Specific chemicals – *continued*

Substance	Hazard	Suitability		Guidance Comments
Hexenes	Highly flammable and irritant	Y9+	(F)	
Hydrazine, anhydrous	Corrosive and toxic	N		Not recommended for use in schools.
Hydrazine hydrate	Toxic and corrosive	N		Not recommended for use in schools.
Hydrazinium salts	Toxic	N		Presumed carcinogenic in humans.
Hydrides (reactive metal)	Highly flammable	T		These materials react with water to give hydrogen.
Hydriodic acid ≥25% ≥10 <25%	Corrosive Irritant	(Y9+)		
Hydrobromic acid ≥40% ≥10 <40%	Corrosive Irritant	(Y9+)		
Hydrochloric acid Concentrated and ≥6.5 M ≥2 M<6.5 M	Corrosive Irritant	(Y9+) (Y7+)	(F)	Do not mix with methanal. Sulphuric acid can be used in place of hydrochloric acid in the preparation of urea/methanal polymer.
Hydrofluoric acid	Very toxic and corrosive	N		Not suitable for use in schools.
Hydrogen Cylinder When prepared as needed Test tube scale	Extremely flammable 	 T T Y7+		Use by staff only. Mixtures between 4% and 74% of hydrogen with air can explode. The ignition temperature is well below red heat and the reaction is catalysed by transition metals and their oxides. Hydrogen is best not dried using sulphuric acid: concentrated calcium chloride solution will do. When hydrogen is being prepared or used, safety screens are essential and eye protection should be worn.
Hydrogen chloride (gas)	Corrosive	(Y9+)	F	
Hydrogen cyanide	Very toxic and highly flammable	N		Not suitable for use in schools.
Hydrogen peroxide ≥71 volume ≥18 volume <71 volume <18 volume	Corrosive Irritant Minimal hazard	T Y12+ (Y7+)		Use by staff only. Reacts violently with propanone, ethanol, glycerol and other organic reducing agents. Finely-divided metals cause violent decomposition.

Table 15.4   Specific chemicals – *continued*

Substance	Hazard	Suitability	Guidance Comments	
Hydrogen sulphide	Highly flammable and very toxic			
Cylinder		N		Not suitable for use in schools.
Gas prepared as needed		Y12+	F	Only small quantities of the gas should be made as needed and only in a fume cupboard. The gas can react violently with metal oxides, peroxides and soda lime.
Solution		Y12+	(F)	
2-hydroxybenzoic acid (salicylic acid)	Harmful	Y9+		Vapour irritates respiratory system.
Hydroxylammonium salts	Irritant	Y12+		
2-hydroxypropanoic acid (lactic acid)	Irritant	Y9+		
Indicators	Most may be harmful. For full details see supplier's data sheet			Indicators should be made up by staff. The weighing is best carried out in a fume cupboard. If the indicators are used in alcohol solution, they should be treated as alcohol from a safety point of view. Care needs to be exercised with old samples of indicators if their exact composition is not known. Treat as a hazard. See *Handbook*.
Solid		T	(F)	
Solution		Y7+		
Iodic(V) acid	Oxidising and corrosive	T		Use by staff only.
Iodine		N		Reactions of solid iodine with ethanal, ammonia and antimony should be avoided as they are potentially explosive. The use of iodine to prepare iodoethane from ethanol and phosphorus is not recommended in schools.
Solid	Harmful	T	F	The reaction with aluminium, catalysed by water, should be carried out in a fume cupboard. The induction period can be reduced if a dilute solution of detergent is used.
Solution ≥1 M	Harmful	(Y7+)	(F)	Use solid under close supervision with students below year 12 if it is to be heated. All heating should be carried out in a fume cupboard unless care is taken to prevent the harmful vapour escaping into the atmosphere.
		Y7+		Use cyclohexane to demonstrate the colour of iodine in solution.
Iodobenzene	Harmful	Y12+	(F)	There is some evidence that this material is explosive above 200°C and samples should not be redistilled. The reaction between potassium iodide and benzenediazonium chloride solution can be carried out provided that the normal precautions are taken for working with diazo compounds.

Table 15.4   Specific chemicals – *continued*

Substance	Hazard	Suitability		Guidance Comments
Iodobutane	Highly flammable and harmful	Y12+	(F)	This substance is sensitive to light. Gloves should be worn.
Iodoethane	Highly flammable and harmful	Y12+	(F)	This substance is sensitive to light. Gloves should be worn.
Iodomethane	Toxic	T	(F)	This substance is a carcinogen and is sensitive to light. Gloves should be worn.
Iodopropane	Flammable and harmful	(Y12+)	(F)	This substance is sensitive to light. Gloves should be worn.

Iron(III) chloride
  Anhydrous solid
  Hydrated solid
  Solution ≥0.7 M
	
	Irritant
	Irritant
		
		Y9+
		(Y7+)
		(Y7+)
				If preparing large volumes of solution, for etching PCBs, beware of dust generated.

Iron(III) nitrate	Oxidising and irritant	Y9+		
Iron(II) sulphate	Harmful	Y7+		
Iron(III) sulphate	Irritant	Y7+		
Iron(II) sulphide	Harmful	Y9+		
Lead	Minimal hazard	Y7+		Wash hands after handling the solid.

Lead salts (general)
  Solid
  Solution ≥0.01M
  ≥0.001 to 0.01M
	
	Toxic
	Toxic
	Harmful
		
		(Y9+)
		(Y9+)
		Y7+
				These salts and solutions are deemed toxic because they may cause harm to the unborn child.

Lead(II) bromide	Toxic	(Y9+)		Use a fume cupboard if the molten solid is to be electrolysed.
Lead(II) chromate(VI)	Toxic	N		The solid is not recommended for use in schools. The production of a precipitate of lead chromate does not present a hazard provided it is not filtered off and dried.
Lead(II) ethanoate (acetate)	Toxic	N		A suspected carcinogen. May cause harm to the unborn child and may impair fertility. Schools should use lead nitrate as a substitute.

Lead(II) nitrate
  Solid
  Solution ≥0.01 M
  ≥0.001 to 0.01 M
	
	Toxic and
	oxidising
	Toxic
	Harmful
		
		(Y9+)
		(Y7+)
				If this material is used to examine the effect of heat on nitrates as a class experiment, very small quantities should be used and only in a well-ventilated laboratory (as the LTEL-TWA for nitrogen dioxide is low).

Table 15.4   **Specific chemicals** – *continued*

Substance	Hazard	Suitability	Guidance Comments
Lead oxides	Toxic	(Y7+)	Vigorous reaction with aluminium, magnesium powders and organic materials.
Lithium	Flammable and corrosive	N T	Do not react lithium with mercury. Lithium is known to explode on heating in air (the reaction being accelerated by moisture). Always use a safety screen. Students should not be allowed to carry out the air reaction. When the material does burn it is difficult to extinguish the flame; use powdered sodium chloride.
		(Y9+)	Students may react the solid with water.
Lithium chloride	Harmful	(Y7+)	
Lithium hydroxide	Corrosive	Y12+	
Lithium nitrate	Oxidising	(Y9+)	Take care; nitrogen dioxide is produced on heating. (See lead(II) nitrate.)
Lithium tetrahydridoaluminate (Lithium aluminium hydride)	Highly flammable	N	This material should be used only in those schools equipped to cope with the problems involved in its use. Sodium borohydride provides a safer alternative as an illustration of reduction in some organic systems.
Magnesium Powder	Highly flammable	N	When mixed with air, the powder can form explosive mixtures. The dust should not be blown into a Bunsen flame. The reactions of magnesium with ammonium dichromate, silver nitrate, sulphur or halogenated hydrocarbons can be unexpectedly violent and should not be carried out in schools. The reaction with alcohol is violent after a long induction period. Magnesium metal is not a suitable alternative for use in the 'thermite' reaction except as a fuse.
		T	The use of magnesium powder to reduce sand to silicon is very exothermic and can be explosive if the sand is not absolutely dry.
Turnings	Highly flammable	(Y9+)	The reaction with acids is very vigorous; careful supervision is needed.
Ribbon	Highly flammable	(Y7+)	If this material is burnt, ensure students do not look directly at the flame.
Magnesium chlorate(VII) (Anhydrone)	Oxidising and irritant	N	Not suitable for use in schools.

Table 15.4   Specific chemicals – *continued*

Substance	Hazard	Suitability		Guidance Comments
Magnesium nitrate	Oxidising	(Y9+)		Take care; nitrogen dioxide is produced on heating. (See lead(II) nitrate.)
Manganese(IV) oxide	Harmful	Y7+		The reaction with potassium chlorate(V) has been the source of many accidents, generally because one or other of the reactants was not pure. Oxygen is better generated from hydrogen peroxide solution using manganese(IV) oxide as a catalyst.
Mercury	Toxic	Y12+	(F)	Use in a well-ventilated room on a spillage tray. The major risk is from long-term exposure to very small quantities of mercury vapour. For this reason the metal should not be left open to the air for very long and never heated except in a very efficient fume cupboard. Anyone wearing gold rings should wear protective gloves! Mercury should not be reacted with ammonia, bromine or chlorine. If mercury is used as the electrode in the electrolysis of sodium chloride solution, the amalgam used should be totally reacted before the mercury is added to the stock of 'dirty' mercury saved for purification. This can be achieved by covering the material with dilute hydrochloric acid and leaving it standing for several days or by the addition of water and an iron nail. The reaction of mercury and its salts with aluminium is very exothermic and difficult to stop. The foil is best left in water and disposed of with other unwanted chemicals.
Mercury salts (general) Solid Solution ≥0.004 M 0.002 to 0.004 M	Very toxic Toxic Harmful	(Y12+) (Y9+) (Y9+)		Cole's modification of Millon's reagent should be used rather than Millon's itself for protein tests. For details of Cole's modification, consult *Hazcards* or *Haz Man*.
Mercury alkyls	Very toxic	N		Not suitable for use in schools.
Mercury(II) chloride	Very toxic	Y12+		
Mercury(II) nitrate	Very toxic	T		Heating this solid produces mercury as well as nitrogen dioxide. It produces mercury fulminate with ethanol.

Table 15.4   Specific chemicals – *continued*

Substance	Hazard	Suitability		Guidance Comments
Mercury(II) oxide	Very toxic	Y12+	(F)	If heated, a fume cupboard is essential.
Methanal (formaldehyde)		N	F	Hydrochloric acid should not be used as the catalyst in the production of condensation polymers involving methanal. Concentrated sulphuric acid (10 M) is suitable as an alternative, the reaction being carried out in a fume cupboard. Although the reaction between hydrogen chloride and methanal is known to form bis-chloromethylether, a potent carcinogen, there is no evidence that any has been detected in schools.
≥25%	Toxic	Y12+		Specimens that have been preserved in methanal solution should be soaked in distilled water for at least an hour and then thoroughly washed especially inside the specimen where methanal may have been trapped. If possible, the methanal should be replaced by a methanal-free preservative after the specimen has been washed. Sealed specimens containing methanal do not present a risk provided they are not opened.
≥1% <25%	Harmful	(Y9+)		
Methanamide (formamide)	Toxic	Y12+		
Methane	Highly flammable	Y7+		Use mains gas supply, not cylinders.
Methanoic acid (formic acid)			F	The reaction with concentrated sulphuric acid produces carbon monoxide.
≥10%	Corrosive	(Y9+)		
≥2% <10%	Irritant	(Y9+)		
Methanol	Highly flammable and toxic	(Y9+)	F	Use under supervision with younger students. Methanol is not generally a suitable alternative to ethanol. The reaction with concentrated sulphuric acid is very exothermic. Use a water bath for all heating activities.
Methoxybenzene (anisole)	Harmful and flammable	Y12+		
Methylamine Liquid and solution ≥20%	Highly flammable and irritant	Y12+	F	Aminobutane is a better substance to use in the study of the reactions of amines.
Methylbenzene (toluene)	Highly flammable and harmful	(Y9+)	(F)	If this material is used to prepare the allotropes of sulphur (dimethyl benzene is a much safer alternative) the liquid should be heated on an electrical heater (or an oil bath which is not heated by a Bunsen burner) in a fume cupboard. Although methylbenzene can be used as a safer alternative to benzene to show electrophilic substitution, methyl benzoate is a much better substitute.

Table 15.4   Specific chemicals – *continued*

Substance	Hazard	Suitability		Guidance Comments
Methyl benzoate	Harmful	Y9+	(F)	This is an excellent substitute for benzene to show electrophilic substitution. Nitration produces mainly methyl 3-nitrobenzoate if the temperature is kept below 15°C.
3-Methylbutanol (iso-amyl alcohol)	Flammable and harmful	Y9+	(F)	
3-Methylbutyl ethanoate (*iso-amyl acetate*)	Flammable	Y12+	(F)	

Methylene blue
  Solid
  Solution
	Irritant	T
		Y7+		

Methyl ethanoate (methyl acetate)	Highly flammable	Y9+	(F)	
Methyl methanoate (methyl formate)	Highly flammable	Y9+	(F)	
Methyl 2-methylpropenoate (methyl methacrylate)	Highly flammable and irritant	T Y12+	F (F)	The substance may cause sensitisation by skin contact. As the vapour is lachrymatory the depolymerisation of Perspex is best carried out in a fume cupboard as a demonstration. The polymerisation reaction initiated with di(dodecanoyl) peroxide is very exothermic if more than a very small quantity of peroxide is used. The material as supplied has a stabiliser added and this needs to be removed by shaking with dilute sodium hydroxide solution before polymerisation is attempted.

Methyl orange
  Solid
  Solution
	Minimal hazard
	Highly flammable
		T
		(Y7+)
				Solvent is an aqueous ethanol mixture.

2-methylpropan-2-ol	Highly flammable and harmful	(Y9+)	(F)	
Methylurea (methylcarbamide)	Harmful	Y12+		
Millon's reagent	Corrosive and toxic	N		As this material contains toxic materials as well as concentrated acid, its use in schools is best avoided. There are alternatives such as Albustix, the Sakaguchi test or Cole's modification of Millon's reagent which are less hazardous. See *Hazcards* or *Haz Man*.
Millon's reagent: Cole's modification	Corrosive and toxic	(Y9+)		Although this material contains toxic material as well as concentrated acid it is safer than traditional Millon's reagent. See *Hazcards* or *Haz Man*.

Table 15.4   Specific chemicals – *continued*

Substance	Hazard	Suitability		Guidance Comments
Molybdenum	Minimal hazard	Y9+		
Naphtha	Flammable and harmful	T	(F)	This material is not the best for storage of alkali metals. Liquid paraffin is much safer!
Naphthalen-1-amine and 2-amine (1-naphthylamine & 2-naphthylamine)	Toxic carcinogenic	X		Banned in schools.
Naphthalene	Harmful	N		The nitration of naphthalene is not a suitable experiment for schools. Naphthalene should not be burnt in the open laboratory.
		Y9+	(F)	This material is not suitable for producing cooling curves unless the heating is carried out with mineral wool plugged loosely in the end of the tube when very little vapour escapes. The experiment is best carried out in a fume cupboard. Alternatives include hexadecan-1-ol and octadecan-1-ol. Phenyl benzoate is a suitable alternative for recrystallisation experiments. Use hot water to melt these solids.
Naphthalen-1-ol and 2-ol (α- and β-naphthol)	Harmful	Y12+		This material should not be nitrated. It is safe to use in the preparation of azo dyes but these should not be isolated as some insoluble azo dyes have been shown to be carcinogenic. However, old samples of these may contain naphthylamine, the presence of which is very hazardous. Only samples of known purity >99% can be considered safe to use for this reaction.
Nessler's reagent	Toxic and corrosive	(Y12+)		The mixture contains potassium iodide, mercury iodide and sodium hydroxide.

Nickel
  powder
  sheet or wire
	Harmful	N
		Y7+
				The dust is a suspected carcinogen and the metal may cause sensitisation by skin contact.

Nickel salts (general) Solid	Harmful	Y12+		These may be carcinogenic if insoluble. The dust is a major problem and care needs to be taken as sensitisation to the material may occur.
Solution ≥ 0.02 M	Harmful	Y9+		These do not seem to present the same problems as the dust but may cause allergic reactions.
Nickel(II) carbonate Solid	Harmful	Y12+		May cause sensitisation by skin contact. If the solid is heated, the residue is toxic nickel(II) oxide. After cooling, this should be dissolved in dilute sulphuric acid with heating.

Table 15.4   **Specific chemicals** – *continued*

Substance	Hazard	Suitability	Guidance Comments
Nickel(II) nitrate	Harmful and oxidising	Y9+	May cause sensitisation by skin contact. Should be treated as a suspected carcinogen, but there should be little problem if dust is not raised, for example, in crystallisation.
Nickel(II) oxide	Toxic	N	This material is often found in use in art departments; it is carcinogenic. The dust should not be inhaled. Its use is not recommended in schools.
Ninhydrin Solid Spray	Harmful Highly flammable	T (Y9+) F	There is no evidence that this substance is carcinogenic. The material is generally supplied as a solution in butan-1-ol and the spray should be used in a fume cupboard with no heat sources near by. Wear gloves.
Nitrates (general)	Oxidising	N Y9+	Nitrates should not be mixed with reducing agents and heated. The reaction with sulphur is not suitable for schools. See also individual nitrates. The action of heat generally produces nitrogen dioxide which is toxic. If, in the thermal decomposition of a nitrate, oxygen is being tested for, care needs to be taken that the splint does not drop into the molten mass.
Nitric acid Fuming 100% Concentrated 70% ≥0.5 M <3 M ≥0.1 M <0.5 M	Corrosive and oxidising Corrosive and oxidising Corrosive Irritant	T (F) (Y9+) (F) (Y7+) Y7+	Any apparatus used should be all glass with no rubber or cork stoppers. If nitrogen dioxide is expected in anything other than the very smallest quantities, a fume cupboard should be used. Nitric acid reacts violently with organic substances, ethanoic acid, ethanol, propanone, aluminium, lithium, magnesium and the alkali metals, thiocyanates and thiosulphates. If crucibles used for heating magnesium are cleaned with nitric acid, then this should be performed in a fume cupboard. The use of mixtures of the acid and alcohol to clean glassware is not safe for school use. If 'aqua regia' is used to dissolve metal alloys, the mixture should be made up fresh, cooled and not stored. The production of a brown coloration in the liquid is a sign that decomposition has started to occur and the mixture should be disposed of by pouring into a very large volume of water.

Table 15.4　Specific chemicals – *continued*

Substance	Hazard	Suitability		Guidance Comments
Nitrites 　Solid	Toxic	Y12+		See ammonium nitrite. Nitrites react with acid eventually to produce toxic nitrogen dioxide. There is potential for the confusion of nitrites with nitrates; they are best stored separately.
Solution	Toxic	Y9+		
Nitrobenzene	Very toxic	Y12+	(F)	There is a danger of cumulative effects if this compound is handed often. Small quantities used once or twice in advanced courses by senior students should not present a danger. Gloves should be worn.
4-nitrobiphenyl	Toxic	X		Banned in schools.
Nitrocellulose	Explosive	N		Not suitable for use in schools.
Nitrogen dioxide	Toxic	N		Hydrogen ignites spontaneously in air if nitrogen dioxide is present. The reactions of the gas with alcohol, hydrocarbons and halogenoalkanes are vigorous and dangerous.
Cylinder		N		Not suitable for use in schools.
Prepared as needed		Y12+	(F)	Small quantities may be prepared in a well-ventilated laboratory under supervision with younger students.
Test-tube quantities		(Y9+)	(F)	Students should be taught to smell gases safely before they are asked to smell nitrogen dioxide.
Nitrogen monoxide	Toxic	N		The gas reacts violently with methanol. The reaction with carbon disulphide is not recommended as explosions have occurred.
Cylinder		N		Not recommended for use in schools.
Prepared as needed		Y12+	(F)	Small quantities may be prepared in a well-ventilated laboratory.
Test-tube quantities		(Y9+)		
Nitrogen triiodide	Explosive	N		Its preparation is not recommended in schools.
Nitromethylbenzenes (nitrotoluenes)	Toxic	N		Not recommended for use in schools.
Nitronaphthalenes	Toxic	X		Banned in schools.
Nitrophenols	Harmful	(Y12+)		These compounds should not be made on a preparative scale. See nitrosophenols.
Nitrophenylamines (nitroanilines)	Toxic	T		These compounds should not be made on a preparative scale.

Table 15.4   Specific chemicals – *continued*

Substance	Hazard	Suitability	Guidance Comments
4-(4-nitrophenylazo) rosourcinol Magneson I 4-(4-nitrophonylaze)-l-napthol Magneson Solution	Minimal hazard (Highly flammable)	T (Y7+)	Indicators should be made up by staff. The weighing is best carried out in a fume cupboard. If the indicators are used in alcohol solution they should be treated as alcohol from a safety point of view. Care needs to be exercised with old samples of indicators if their exact composition is not known. Treat as a hazard. See *Handbook*.
Nitrosoamines (general)	Toxic	X	Banned in schools.
Nitrosnaphthalenes	Toxic	X	Banned in schools.
Nitrosophenols, 2- and 3-isomers	Toxic	X	Banned in Schools. These could be unwittingly prepared during the nitration of phenol. For this reason the nitration of phenol should not be carried out with the usual nitrating mixture but with a solution of sodium nitrate and sulphuric acid in an ice bath. The products of the reaction should only be handled with gloves on and the reaction tube wapped in paper and discarded at the end of the experiment. Do not attempt to wash out the tubes.
4-nitrosophenol	Toxic	(Y12+)	This material is not thought to be carcinogenic unlike other nitrosophenols.
Nitrosophenols, 2- and 3-isomers	Toxic	X	Banned in Schools. These could be unwittingly prepared during the nitration of phenol. For this reason the nitration of phenol should not be carried out with the usual nitrating mixture but with a solution of sodium nitrate and sulphuric acid in an ice bath. The products of the reaction should only be handled with gloves on and the reaction tube wapped in paper and discarded at the end of the experiment. Do not attempt to wash out the tubes.
4-nitrosophenol	Toxic	(Y12+)	This material is not thought to be carcinogenic unlike other nitrosophenols.
Octane	Highly flammable	Y9+ (F)	
Oct-1-ene	Highly flammable and harmful	Y12+ (F)	
Oleum	Very corrosive	N	Not recommended for use in schools.
Osmic acid	Very toxic	N	Not recommended for use in schools.

Table 15.4   Specific chemicals – *continued*

Substance	Hazard	Suitability		Guidance Comments
Oxygen	Oxidising	N		The preparation of oxygen from potassium chlorate(V) should be avoided. The use of hydrogen peroxide and manganese(IV) oxide is much safer.
Cylinder		T		
Prepared as needed		Y7+		
'Oxygen mixture' (potassium chlorate(V) plus manganese(IV) oxide	Oxidising and harmful	N		Not recommended for use in schools for the preparation of oxygen, but the reaction is acceptable on a small scale to demonstrate catalytic decomposition. Oxygen is best prepared by the action of manganese(IV) oxide on hydrogen peroxide solution. See also potassium chlorate(V).
Paraffin (Kerosene)	Flammable	(Y9+)		This is the paraffin that is used as a fuel.
Paraffin Liquid (medicinal paraffin)	Minimal hazard	Y7+		Used in 'cracking hydrocarbon' experiments.
Paraquat	Toxic	N		Liquid not suitable for use in schools; material supplied for professional agricultural use is particularly dangerous. Some pesticides may still contain paraquat but formulations now available are very dilute, with the chemical incorporated into soluble granules rather than supplied as a liquid.
Pentane	Extremely flammable	Y9+	(F)	All sources of ignition should be extinguished.
Pentan-1-ol and -2-ol	Flammable and harmful	Y9+	(F)	
Pentan-2-one	Highly flammable and harmful	Y9+	(F)	
Pentan-3-one	Highly flammable	Y9+	(F)	
Pentyl ethanoate (amyl acetate)	Flammable	Y9+	(F)	
Petrol	Highly flammable	X		Must not be used except in motors as it contains more than 0.1% benzene(2).
Petroleum spirits (pet. ether)				
B.pt. <80°C	Higly flammable and harmful	Y12+	(F)	Small quantities may be used in a well-ventilated laboratory provided no sources of ignition are present. Contains hexane. Use 80-100°C petroleum spirit unless low b.pt. material is essential.
B.pt. >80°C	Highly flammable	Y9+	(F)	Small quantities may be used in a well-ventilated laboratory provided no sources of ignition are present.

Table 15.4   Specific chemicals – *continued*

Substance	Hazard	Suitability		Guidance Comments

Phenol
  Solid
  Solution ≥2.5%
  ≥3% <25%
	Toxic
	Toxic
	Harmful
		(Y12+)
		Y12+
		Y12+
				Use sulphuric acid to prepare methanal/phenol polymers. Do not nitrate with a sulphuric/nitric acid mixture; use a mixture of sodium nitrate and acid or 4 M nitric acid alone. Gloves need to be worn if the solid is used, as phenol can cause burns if it comes into contact with the skin. Glycerol or preferably polyethylene glycol rubbed into the affected area can prevent blistering.

Phenolphthalein
  Solid
  Solution in ethanol
	Minimal hazard
	Highly flammable
		T
		Y7+
				(Laxative)
				The ethanol provides the major hazard.

Phenylamine (aniline) and salts
  Liquid solid
  Solution ≥1%
  ≥0.1 <1%
	Toxic
	Toxic
	Harmful
		(Y12+)
		Y12+
		Y12+
			F	Phenylamine reacts violently with nitric acid and peroxides. See the sections on naphthalene-1-and-2-ol for notes on the preparation of azo dyes. When diazonium compounds are prepared it is important to check that the reaction is complete by adding extra nitrite solution, and testing a sample with acidified potassium iodide solution.

Phenylammonium salts	Toxic	Y12+	(F)	These are much easier to handle and safer than the liquids.
Phenyl benzoate	Flammable	Y12+		This material is a substitute for benzene in nitration reactions and can be used as an example of recrystallisation using ethanol as the solvent.
Phenylethanone	Harmful	Y12+	(F)	
Phenylethene (styrene)	Flammable and harmful	Y12+	F	The polymerisation should be initiated with di(dodecanoyl) peroxide. The material as supplied has a stabiliser added and this needs to be removed by shaking with dilute sodium hydroxide solution before polymerisation is attempted.
Phenylhydrazine and salts	Toxic	Y12		Avoid skin contact.
Phenyl 2-hydroxybenzoate (phenyl salicylate, salol)	Minimal hazard	Y7+		Used for melting point determinations and differences in crystal growth when cooled at different rates.
Phenyl thiocarbamide (PTC) (Phenylthiourea) (PTU)	Toxic	(Y9+)		For testing the ability to taste this substance, use only a paper strip. The strip should contain no more than 0.1 mg. Stock solutions and solid should be locked away and never tasted.
Phosphides (metal)		N		Not recommended for use in schools, but if kept, use a rubber stopper.

Table 15.4　　Specific chemicals – *continued*

Substance	Hazard	Suitability		Guidance Comments
Phosphine	Toxic	T	F	The preparation of the gas should be carried out in a fume cupboard. Impurities in the gas render the gas spontaneously flammable in air.
Phosphoric(V) acid 　Liquid 　≥2.5 M 　≥0.5 M <2.5 M	Corrosive Corrosive Irritant	Y12+ Y9+ Y9+		
Phosphorus (red)	Highly flammable	Y12+	(F)	Small quantities may be used in a well-ventilated laboratory. Forms explosive mixtures with chlorates and metal oxides. All spatulas etc. that have been used to handle red phosphorus need to be washed with water, the washings filtered and the filter paper burnt in a fume cupboard.
Phosphorus (white)	Highly flammable and toxic	T	F	Liable to ignite spontaneously; always handle with tongs and cut under water in a strong container such as a mortar. Always have a 0.2 M solution of copper(II) sulphate ready to remove phosphorus from skin or clothes. For combustion reactions, ignite with a hot glass rod. White phosphorus ignites on contact with iodine but explodes with liquid bromine. In the reaction with chlorine no heat is needed and leaving the solid wet allows time for the manipulation of the apparatus before the phosphorus ignites.
Phosphorus(V) oxide	Corrosive	N		The oxide should not be used for dehydration of methanoic acid as large volumes of carbon monoxide are produced.
		(Y12+)		If a crust has formed on the solid it should be removed and treated as a solid acid for disposal (unreacted oxide will be sticking to the solid). The oxide reacts violently with water and iodides. Sodium and potassium metals ignite on contact with phosphorus(V) oxide.
Phosphorus pentabromide	Corrosive	(Y12+)	F	Vigorous reaction with water.
Phosphorus pentachloride	Corrosive	N		Violent reaction with sodium and potassium.
		(Y12+)	(F)	Small quantities may be used in a well-ventilated laboratory to investigate the action with alcohols but always add the chloride to the alcohol. For the preparation of acid chlorides, always lead any gases produced down into a sink with the tap running and distil the mixture from a water bath, not with direct heat.
Phosphorus tribromide	Corrosive	(Y12+)	F	Vigorous reaction with water.

Table 15.4   Specific chemicals – *continued*

Substance	Hazard	Suitability		Guidance Comments
Phosphorus trichloride	Corrosive	N		Violent reaction with sodium and potassium.
		(Y12+)	(F)	Small quantities may be used in a well-ventilated laboratory to investigate the action with alcohols but always add the chloride to the alcohol. For the preparation of acid chlorides, always lead any gases produced down into a sink with the tap running and distil the mixture from a water bath, not with direct heat.
Phosphorus trichloride oxide (oxychloride)	Corrosive	(Y12+)	F	
Potassium	Highly flammable and corrosive	N		Potassium should never be burnt in pure oxygen or chlorine. Potassium reacts explosively with ice. Alloys of potassium with sodium should not be prepared as they spontaneously inflame in air. The reaction with 1,1,1-trichloroethane and other halogenated hydrocarbons is dangerous.
		T		Schools are advised not to keep large quantities of the metal and stocks should be checked on a regular basis. Old stocks of this metal that have started to go yellow due to the production of superoxide may explode when cut. It should be destroyed by adding it, a small piece at a time, to 2-methylpropan-2-ol.
Potassium amide	Corrosive	N		Not recommended for use in schools.
Potassium bromate(V)	Oxidising and toxic	Y12+		This material is carcinogenic. Sodium bromate(V) does not appear to have the same hazardous properties, it is harmful and oxidising.
Potassium carbonate	Irritant	Y7+		

Table 15.4 Specific chemicals – *continued*

Substance	Hazard	Suitability	Guidance Comments
Potassium chlorate(V)	Oxidising and harmful	X	The making of explosives without a licence is forbidden by law. This material must never be mixed with:
		X	(i) sulphur or sulphides,
		N	(ii) ammonium salts,
		N	(iii) phosphorus,
		N	(iv) 2,4,6-trinitrophenol (picric acid),
		N	(v) fine metal powders such as aluminium and magnesium,
		N	(vi) 3,4,5-trihydroxybenzoic acid (gallic acid).
		N	Mixtures with fuels such as sugar and hydrocarbons are dangerous.
		N	The reaction with concentrated sulphuric acid produces unstable chlorine dioxide which has a serious risk of explosion.
		Y12+	Solutions of potassium chlorate produced during experiments to study the effect of temperature change on solubility should not be allowed to dry out on clothing etc. The solid may be recrystallised at the end of the experiment but this solid should not then be used for heating experiments and it may be considered safer to dispose of the material.
Potassium chlorate(VII)	Oxidising and harmful	Y12+	This material is produced in the disproportionation reaction of potassium chlorate(V). If the solid is crystallised out it should not be stored for use in other experiments.
Potassium chromate(VI) Solid and solution ≥0.025 M	Irritant	N	May cause sensitisation by skin contact. The reactions with aluminium, magnesium, carbon, sulphur and phosphorus are dangerous. Chromic acid made by the action of concentrated sulphuric acid on the solid chromate(VI) is not recommended for use in schools.
		Y12+	The oxidation of ethanol under reflux has been the source of many accidents. In preparative exercises the acidified chromate solution (no solid undissolved) should be added dropwise with thorough mixing or stirring. The temperature of the mixture should be raised very slowly before refluxing.
		Y9+	Test tube scale.
Potassium cyanide	Very toxic	N	Not recommended for use in schools. See entry for cyanides.
Potassium dichromate(VI)	Irritant	Y9+	See potassium chromate(VI).
Potassium fluoride Solid and solutions ≥4 M ≥0.5 M < 4M	Toxic Harmful	T Y12+	

Table 15.4   **Specific chemicals** – *continued*

Substance	Hazard	Suitability	Guidance Comments

Potassium hexacyanoferrate(II)
  Solid
  Solution
	Minimal hazard	Y12+
		Y9+
			Forms explosive mixtures with nitrites and gives hydrogen cyanide with concentrated acids. Do not heat the solid.

Potassium hexacyano-
ferrate(III)
  Solid
  Solution
	Minimal hazard	Y12+
		Y9+
			May explode with ammonia, forms explosive mixtures with nitrites and hydrogen cyanide with concentrated acids. Do not heat the solid.

Potassium hydrogensulphate
  Solid
  Solution
	Corrosive	Y9+
		Y7+
			Treat solution as dilute sulphuric acid.

Potassium hydroxide
  Solid or melt
  Concentrated solution
  ≥0.5 M
  ≥0.05 M <0.5 M
	Corrosive
	Corrosive
	Corrosive
	Irritant
		(Y12+)
		(Y12+)
		(Y9+)
		(Y7+)
			Eye protection should be worn at all times when using this material, no matter how dilute the solution involved.

Potassium iodate(V)	Oxidising	Y12+	Reacts violently with aluminium, magnesium, carbon, sulphur and phosphorus. Potassium iodate has also been known to react violently with disulphate(IV) (metabisulphite) if the dry solids are mixed and water then added. The reaction with concentrated sulphuric acid can be carried out on a test-tube scale, the residue being disposed of by mixing with a large volume of water.
Potassium manganate(VII) (permanganate)	Oxidising and harmful	N	Manganate(VII) forms an explosive mixture with concentrated sulphuric acid or phosphoric acid. Manganate(VII) reacts violently with metal powders and combustible materials. Ammonia and ammonium compounds may produce explosive salts when mixed dry.
		T (F)	Mixtures of the solid and propane-1,2,3-triol spontaneously inflame.
		(Y12+)	If the solid is being used to prepare chlorine gas it is essential that the acid used is checked and that the solid is covered with water. It is best not to use sulphuric acid for drying the gas.
		(Y7+)	If the solid is heated in a tube, a very fine dust is produced. Use a loose-fitting cermaic wool plug or a fume cupboard. While acidified dilute hydrogen peroxide can be used to clean stains left by manganate(VII), concentrated hydrogen peroxide should not be used.

Table 15.4　Specific chemicals – *continued*

Substance	Hazard	Suitability		Guidance Comments
Potassium nitrate(V)	Oxidising	N		Nitrates form explosive mixtures with aluminium, magnesium, sodium, potassium and other metals plus ammonium salts, cyanides, sulphides, thiosulphates and ethanoates.
		(Y7+)		Use under supervision with **all** students. If, in the thermal decomposition of a nitrate, oxygen is being tested for, care needs to be taken that the splint does not drop into the molten mass.
Potassium nitrite	Oxidising and toxic	N		It should not be reacted with phenol which forms an explosive mixture. The solid is dangerous with ammonium salts, cyanides and thiosulphates.
		Y12+		Reacts with acids to produce nitrogen monoxide which gives toxic nitrogen dioxide in air.
Potassium peroxodisulphate(VI)	Oxidising and harmful	Y9+		Short shelf-life.
Potassium phosphate	Minimal hazard	Y7+		
Potassium sulphide	Corrosive	(Y9+)	F	Produces toxic hydrogen sulphide with acids.
Potassium thiocyanate	Harmful	N		Dangerous with concentrated sulphuric acid when the gas carbonyl sulphide is produced. Boiling with dilute acid has the same effect but very slowly.
		Y9+		The very dilute solution provides very little hazard if used in the test for iron(III) but the material should not be evaporated to dryness.
Procion Yellow Solid Solution	Irritant Minimal hazard	T (Y7+)	(F)	Sensitiser. Staff should make solutions for reactions in a fume cupboard. Solutions are safe for all levels.
Propanal	Highly flammable and irritant	Y9+	(F)	
Propane Cylinder	Extremely flammable	N		Not recommended for use in school except if there is a piped gas supply.
Propanoic acid	Corrosive	Y9+	(F)	
Propan-1-ol and 2-ol	Highly flammable	Y9+	(F)	Always use a water bath for heating.
Propanone	Highly flammable	Y7+	(F)	The reaction with halogenated hydrocarbons such as trichloromethane can be vigorous after a long induction period. (Ethyl ethanoate provides a safer alternative for measurement of intermolecular forces.) Nitric acid reacts violently with propanone. All heating should be done on a water bath.

Table 15.4   Specific chemicals – *continued*

Substance	Hazard	Suitability		Guidance Comments
Propenamide (acrylamide)	Toxic	N		Polyacrylamide gels for electrophoresis of DNA are best purchased as the polymer.

Propylamine
  Liquid
 	
  Solution
	Highly flammable and corrosive	Y12+
		
		Y9+
			F	Butylamine provides a safer alternative.

Propyl ethanoate	Highly flammable	Y9+	(F)	
Pyridine	Highly flammable and harmful	Y12+	F	No naked flames should be present.
Quinine	Harmful	T		Cold tea provides a good alternative for taste-bud tests.
Resazurin	Minimal hazard	Y7+		Available in tablet form.
Saponin	Irritant	Y9+		
Screened methyl orange	Minimal hazard	Y7+		See indicators.
Selenium and compounds	Toxic	N		An exhibition sample may be kept.
Silicon tetrachloride	Irritant	T (Y12+)	F	There is much evidence of this material presenting a hazard in school use. Unopened bottles have exploded in stores due to a build up of hydrogen chloride gas during manufacture. With bottles that have been opened, the major problem seems to be the stopper being covered with the chloride or absorbing it in the case of rubber; the chloride then reacts with moisture in the air to produce a solid block of silica or fuses a glass stopper into the neck of the bottle. This can mean undue force is needed to open the bottle with the consequent risk of the bottle shattering and the rapid evolution of gas. Bottles of silicon tetrachloride should therefore be opened in a fume cupboard while wearing gloves and with the bottle covered with a dry cloth. It is best not to use the liquid on very wet days. Only small quantities of the liquid should be stored. If possible, the purchase of ampoules is a much safer alternative. Once liquid has been removed from the bottle it should not be returned. Reacts vigorously with sodium and other reactive metals. Reacts vigorously with water.

Table 15.4   Specific chemicals – *continued*

Substance	Hazard	Suitability	Guidance Comments
Silver nitrate Solid	Corrosive and oxidising	(Y12+)	Reacts with ammonia to form explosive products under certain conditions. A mixture with magnesium explodes in the presence of water. The reaction with ethanol is very vigorous. See also Tollen's reagent.
Solution ≥0.5 M	Corrosive	Y12+	
≥0.2 M < 0.5 M	Irritant	(Y7+)	
Soda lime	Corrosive	(Y7+)	
Sodamide	Corrosive and flammable	N	Not recommended for use in schools.
Sodium	Highly flammable and corrosive	N	Reacts explosively with acids. Forms explosive mixtures with poly-halogenated hydrocarbons. Violent reactions with bromine, iodine, mercury and sulphur. The addition of water to sodium (rather than the usual reaction) is explosive due to the production of sodium oxide and sodium hydride and is much too dangerous to be tried in schools. Alloys of sodium with potassium spontaneously inflame in air.
		T	In combustion reactions with oxygen and chlorine, only small pieces should be used. In the reaction with water only a small piece of sodium should be used otherwise the hydrogen will ignite. This will also occur if the sodium is placed on a piece of filter paper or the water is above 40°C. The sodium should not be trapped in an attempt to collect the hydrogen unless an apparatus specifically designed for the purpose is used.
		(Y12+)	For the sodium fusion test, use a very small amount (0.2 g) of sodium with 0.1 g of solid and a safety screen.
Sodium amalgam	Toxic	Y12+	If mercury is used as the electrode in the electrolysis of sodium chloride solution, the amalgam produced should be totally reacted before the mercury is added to the stock of dirty mercury saved for purification. This can be achieved by covering the material with dilute hydrochloric acid and leaving to stand for several days or by adding water and an iron nail.
Sodium azide	Toxic	N	Contact with acid liberates a very toxic gas.
Sodium borohydride	Toxic and flammable	Y12+	
Sodium carbonate	Irritant	Y7+	

Table 15.4   **Specific chemicals** – *continued*

Substance	Hazard	Suitability	Guidance Comments
Sodium chlorate(I) (hypochlorite) ≥10% available chlorine ≥5% <10% available chlorine	 Corrosive Irritant	 (Y9+) (Y7+)	Dangerous with concentrated sulphuric acid. Forms explosive products with ammonium salts, methanol and amines. Gives off chlorine with acids. If used to prepare chlorine, 5 M HCL should be used. This material is unstable in sunlight and should be stored in a cupboard. As the solution gives off oxygen during storage it should be assumed that there will be a pressure release on opening. The problem is worse in summer and accelerated by some transition metal compounds.
Sodium chlorate(V)	Oxidising and harmful	Y12+	See potassium chlorate for guidance.
Sodium chromate(VI) Solid and solution ≥0.025 M		N Y12+ Y9+	May cause sensitisation by skin contact. The reactions with aluminium, magnesium, carbon, sulphur and phosphorus are dangerous. Chromic acid made by the action of concentrated sulphuric acid on the solid chromate(VI) is not recommended for use in schools. The oxidation of ethanol under reflux has been the source of many accidents. In preparative exercises the acidified chromate solution (no solid undissolved) should be added dropwise with thorough mixing or stirring. The temperature of the mixture should be raised very slowly before refluxing. Test tube scale.
Sodium dichromate(VI)	Irritant	Y9+	See sodium chromate(VI).
Sodium dithionite	Harmful	(Y9+)	Supply the material in solution for class use. Weigh out in a dry beaker. Sulphur dioxide gas may be produced.
Sodium dodecyl sulphate (lauryl sulphate)	Harmful	Y9+	
Sodium ethanedioate (oxalate) Solid Solution ≥0.3 M	 Harmful Harmful	 Y12+ Y9+	
Sodium fluoride Solid and solutions ≥4 M ≥0.5 M < 4 M	 Toxic Harmful	 T Y12+	
Sodium hexanitrocobaltate(III)	Oxidising and toxic	Y12+	Starts fires with combustible material.
Sodium hydride	Flammable	T	Gives off hydrogen with water.
Sodium hydrogensulphate Solid Solution	 Corrosive	 Y9 Y7+	 Treat solution as dilute sulphuric acid.

Table 15.4   Specific chemicals – *continued*

Substance	Hazard	Suitability		Guidance Comments
Sodium hydrogensulphite	Harmful	(Y7+)		
Sodium hydroxide 　Solid or melt 　Concentrated solution 　≥0.5M 　≥0.05M<0.5M	 Corrosive Corrosive Corrosive Irritant	 (Y12+) (Y12+) (Y9+) (Y7+)		Eye protection should be worn at all times when using this material, no matter how dilute the solution involved.
Sodium metabisulphite	Harmful	(Y7+)	(F)	
Sodium nitrate	Oxidising	N (Y7+)		Use under supervision with all students. Nitrates form explosive mixtures with aluminium, magnesium, sodium, potassium and other metals plus ammonium salts, cyanides, sulphides, thiosulphates and ethanoates. If, in the thermal decomposition of a nitrate, oxygen is being tested for, care needs to be taken that the splint does not drop into the molten mass.
Sodium nitrite	Oxidising and toxic	N Y12+		It should not be reacted with phenol which forms an explosive mixture. The solid is dangerous with ammonium salts, cyanides and thiosulphates. Reacts with acids to produce nitrogen monoxide which gives toxic nitrogen dioxide in air.
Sodium orthovanadate	Toxic	Y12+		Ammonium metavanadate provides a much cheaper alternative.
Sodium pentacyanonitrosylferrate(II) (nitroprusside)	Very toxic	(Y12+)		Toxic by inhalation and ingestion. Do not decompose by heat.
Sodium peroxide	Oxidising and corrosive	(Y12+)		Combustible materials may ignite spontaneously especially if damp. The reactions with ethanoic acid and enthanoic anhydride are explosive. It forms an explosive mixture with tin(II) chloride. Reacts with water to give sodium hydroxide.
Sodium sulphide 　Solid 　Solution	 Corrosive	 Y12+ (Y9+)	 F	Reacts with acid to give off toxic hydrogen sulphide.
Strontium	Flammable	T		Handle with care.
Strontium salts				Similar to the corresponding calcium salts.
Sulphides of heavy metals		(Y9+)	(F)	Use a fume cupboard if any reactions are to be investigated involving the use of acid or if roasting an ore as toxic gases are produced.

Table 15.4   **Specific chemicals –** *continued*

Substance	Hazard	Suitability		Guidance Comments
Sulphur	Flammable	X	(F)	The reaction of sulphur with oxidising agents such as chlorates must not be carried out in schools.
		N		The reactions with magnesium, aluminium or more reactive metals are too violent and should not be carried out.
		T		The reaction with zinc is possible provided small amounts are used, the mixture is not restricted in any way and safety screens are used.
		(Y7+)		The reactions with copper and iron may be carried out. Sulphur burns to give off the toxic gas sulphur dioxide; all reactions that involve the possibility of sulphur igniting should be carried out in a fume cupboard.
Sulphur chlorides	Corrosive	(Y12+)	F	Reacts with water producing a range of sulphur compounds. The use of a fume cupboard is essential and supervision provided if students use the material.
Sulphur dichloride dioxide (sulphuryl chloride)	Corrosive	(Y12+)	F	
Sulphur dichloride oxide (thionyl chloride)	Corrosive	(Y12+)	F	Irritating to the respiratory system. Wear gloves.
Sulphur dioxide	Toxic			Many students may be susceptible to asthma attacks which can be brought on by small quantities of the gas. A warning should be issued before the experiment whenever there is a possibility of the gas being evolved in a reaction.
Canister		(Y12+)	F	The valves of canisters of sulphur dioxide should be checked and replaced regularly as they are prone to corrode. The valve should not be screwed down too tightly as this eventually produces a hole too large for the valve to seal.
Preparation: large scale		(Y12+)	F	Rubber tubing that has been used for reactions involving the gas should be discarded after use.
Preparation: test tube		(Y9+)	(F)	
Aqueous solution		(Y7+)	(F)	
Sulphuric acid Concentrated	Corrosive	N		The reactions with chlorates(V) and manganate(VII) produce spontaneously explosive products.
		T		The reaction with sugar can produce large volumes of carbon monoxide and the product needs to be washed thoroughly before it is touched. The reaction with water is very exothermic and solutions should be made by the addition of the acid to the water, and never the other way round.
≥1.5 M	Corrosive	(Y9+)		
≥0.5 < 1.5 M	Irritant	(Y7+)		

Table 15.4 Specific chemicals – *continued*

Substance	Hazard	Suitability		Guidance Comments
Tellurium	Harmful	N		Keep as exhibition sample only.
Tellurium compounds	Harmful	N		Not suitable for use in schools.
Tetrachloromethane	Toxic	N (Y12+)	F	It should not be used as a solvent or in sodium fusion reactions. Reactions with magnesium and alumimium are violent. Use only in an efficient fume cupboard to show its unique properties.
Thallium metal and salts	Toxic	N		Not suitable for use in schools.
Thiocyanates Solid Solution	Harmful	Y12+ Y9+		Dangerous with concentrated sulphuric acid when the gas carbonyl sulphide is produced. Dilute acid with boiling has the same effect but very slowly. The very dilute solution provides very little hazard if used in the test for iron(III) but the material should not be evaporated to dryness.
Thiourea (thiocarbamide)	Harmful	Y12+		
Thorium salts	Radioactive and irritant	(Y12+)		These may only be kept by schools authorised to do so under DFE AM 1/92 working in categories A and B except for small quantities in thoron generators (not Scotland).
Tin(II) chloride	Irritant	Y9+		Forms explosive mixture with oxidising agents such as nitrates and peroxides.
Tin(IV) chloride	Corrosive	Y12+	F	Reacts with water to give hydrogen chloride. Forms an explosive mixture with turpentine.
Titanium tetrachloride	Corrosive	T	F	Violent reaction with water producing hydrogen chloride. Bottles of this material should be opened with caution, covering with a dry cloth as hydrogen chloride gas may escape under pressure.
Tollen's reagent	Corrosive	(Y9+)		Use a clean test tube which should be heated on a water bath only. Do not allow to boil dry as the solid product is explosive. Discard any residues into large volumes of water and wash the tube out with dilute nitric acid. Make only as needed, do not store the solution and do not add excess solution or product of reaction to silver residue bottles.

Table 15.4   **Specific chemicals** – *continued*

Substance	Hazard	Suitability		Guidance Comments
1,1,1-trichloroethane	Harmful	(Y9+)	(F)	Small quantities may be used in a well-ventilated laboratory. This liquid is a safer solvent in situations where tetrachloromethane or trichloromethane might have been used since the hazards are less severe. It should nevertheless be treated with care. The reactions with metals such as magnesium, aluminium and sodium are violent or explosive. The manufacture of this material has been phased out as it is an ozone-depleter and is no longer available.
2,2,2-trichloroethanediol (chloral hydrate)	Toxic	Y12+	F	
Trichloroethanoic acid	Corrosive	Y12+		Do not heat; decomposition can produce a toxic gas.
Trichlorethene (trichloroethylene)	Harmful	Y12+	(F)	This liquid is a safer solvent in situations where tetrachloromethane or trichloromethane might have been used since the hazards are less severe. It should nevertheless be treated with care. The reactions with metals such as magnesium, aluminium and sodium are violent or explosive. Vigorous reaction with alkalis.
Trichloromethane (chloroform)	Toxic	N		Chloroform should never be heated to a high temperature in air as thermal decomposition can cause the production of carbonyl chloride (phosgene). The reactions with aluminium and magnesium are violent; those with sodium, potassium and lithium are explosive. Alkalis in the presence of alcohols or ketones show violent reactions.
		(Y12+)	F	This material should only be used in an efficient fume cupboard where there is no suitable alternative available and never as a solvent. The reaction with propanone to show intermolecular forces may become violent a long time after mixing if traces of base are present. Ethyl ethanoate can be used in place of the propanone. Dichloromethane is a suitable alternative for extraction of caffeine.
Trimethylamine	Highly flammable and irritant	(Y12+)	F	
3,4,5-trihydroxybenzoic acid (gallic acid)	Irritant	Y12+		

Table 15.4   Specific chemicals – *continued*

Substance	Hazard	Suitability		Guidance Comments
2.4.6-trinitrophenol (picric acid)	Explosive and toxic	N T		Reactions of this material with alkali and chlorates should not be attempted. This material is explosive and if stocks of the material need to be kept in schools they should be checked regularly to ensure the material is not dry. Stocks should be kept damp.
Solution		Y12+		The material should only be used in solution.
Triiodomethane (iodoform)	Harmful	Y12+		This is usually encountered in the iodoform test for activated methyl groups. Care needs to be exercised with the reagents involved in the test.
Turpentine	Flammable and harmful	T	F	The reaction with chlorine should be carried out in a fume cupboard. Some varieties are potent sensitisers.
Uranium	Toxic	N		Not suitable for use in schools.
Uranium salts Solid Solution	Toxic Toxic	T Y12+		100 g of uranium compounds may be kept by schools authorised under category C of the DES AM 1/92. Those working in categories A and B may hold more. Solutions used for testing for sodium should be kept in a locked cupboard along with other uranium salts.
Uranyl(VI) zinc ethanoate	Toxic	Y12+		Only use in solution as test for sodium and dispose of the result with a large excess of water. See entry above for storage.
Vanadium(V) oxide	Harmful	Y12+		If this material is being used to demonstrate the contact process, ensure no dust escapes.
Wijs solution	Corrosive and flammable	(Y12+)	(F)	
Xylene cyanol Solid Solution		Y12+ Y9+		See notes on indicators.
Zinc (powder)	Flammable	N T (Y7+)		The dust reacts violently with sulphur, iodine, manganese(IV) oxide and potassium chlorate(V). For reactions with sulphur and iodine use coarse, not fine, powder. Suitable for reactions with dilute acids, salts solutions and copper oxide.
Zinc chloride (anhydrous)	Corrosive	Y12+		
Zinc chromate(VI)	Toxic	N		Solid not suitable for use in schools. Could be made by precipitation from solution, as long as the product is not isolated.

Table 15.4   Specific chemicals – *continued*

Substance	Hazard	Suitability	Guidance Comments
Zinc iodide (anhydrous)	Corrosive	Y12+	
Zinc nitrate	Oxidising	Y9+	
Zinc sulphide	Harmful	Y9+	
Zinc sulphate	Irritant	Y7+	

16 USING ELECTRICITY

16.1   General safety considerations

When carrying out practical activities involving electricity the primary objective should be to ensure both teacher and pupil safety by minimising their exposure to "live" parts of the apparatus, regardless of the voltage of the electrical current. Teachers should be fully familiar with the equipment before experiments are carried out. To reduce the risk of electric shock, electricity requires safe usage and safe equipment. This is only partly achieved by the use of automatic aids such as residual current devices (RCDs) and other earth eakage detectors. All staff must know how to deal with not only the victim of an electric shock but also the hazards of the circumstances surrounding it.

One source of electrical hazards might be appliances of poor quality. DIY installations or equipment borrowed from home or other sources might also create hazards.

The Electricity at Work Regulations 1989, made under the HSWA, cover the provision and maintenance of electrical equipment in the workplace. The Regulations place a duty on the employer to ensure that electrical equipment and installations are in safe working order. Although the main duty is on the employer, the employee also has duties under these Regulations, in particular, to co-operate with their employer in meeting the requirements of the Regulations.

Under the Provision and Use of Work Equipment Regulations 1992, equipment must be suitable for its purpose. Equipment intended for the domestic market may well not meet the more stringent standards required for schools. Schools are therefore advised to purchase equipment from reputableeducational suppliers who should guarantee its suitability. Brief instruction may be needed before certain equipment is used by anyone unfamiliar with it.

Table 16.1   Cells and batteries

Equipment or Procedure	Hazard(s)	Suitability	Guidance Comments	Reference
Disposable cells button cells	Pupils may put small cells in their mouths and swallow them		Button cells are used in stopwatches, calculators, etc. If they are used, care must be taken to ensure that cells are not removed from their holders.	*Handbook Safeguards*
lithium cells	Short circuiting may cause serious overheating as might attempts to recharge or mix with other types	N	Not suitable for pupil use except where fitted inside commercial equipment.	
	Disposal in a fire may cause an explosion			
zinc carbon	Leakage of electrolyte, especially if kept for too long or left in place when stored flat	Y7+	The teacher should recognise the limited shelf life of these cells and organise a system of stock control.	
		T	Pupils should be made aware of good practice and the wide variety of other cells available. With care these cells can be cut open to show the component parts.	
zinc chloride		Y7+	Zinc chloride cells may be recharged with certain charging units but these cells quickly lose their capacity and cannot be considered truly re-chargeable.	
manganese alkaline	Overheating on short-circuit	Y7+	The lower resistance may be useful in certain applications.	
Rechargeable cells Nicad	There is a small risk of explosion with some types if they are excessively over-charged	Y7+	Pupils may use these cells but should not be allowed to recharge them and should be warned against rapid discharge.	*Handbook Safeguards*
		T	Use the correct regulator and organise a recharging procedure. See note on zinc chloride cells.	
other rechargeables, for example gel filled	Corrosive electrolyte in 'dryfit' and other similar lead acid cells	T	No attempt should be made to break the seals on these cells.	
short circuiting	Overheating. [Poor storage may cause accidental short circuit]	N	Keep connecting wires and conductors separate from cells when stored.	

Table 16.1   Cells and batteries - *continued*

Equipment or Procedure	Hazard(s)	Suitability	Guidance / Comments	Reference

Wet cells
  lead acid
  car battery
  NiFe cells
	Spills of corrosive electrolyte. Production of explosive gases when charging	T	The teacher or technician should control recharging procedures.	*Handbook*
				Safeguards

		(Y12+)	Class use should be restricted to older pupils but younger pupils may work with small lead plates in dilute sulphuric acid to show basic principles.	
short circuiting	Injury if dropped Melting or ignition of components	T N	Pupils should not carry large cells. Never allow deliberate short circuiting and store cells to avoid accidents happening.	
other voltaic or wet cells	If the cells are from an unknown source or are obtained by industrial donation their contents and quality may not be suitable	N	Unless the quality and type of electrolyte is known, such donations should not be accepted.	

Table 16.2   Power supplies

Equipment or Procedure	Hazard(s)	Suitability	Guidance / Comments	Reference
low voltage up to 25 V (LV) for basic work	Electric shock if the circuit work involves large inductance	(Y7+)	Strictly limited to less than 25 V and currents below 10 A.	*Handbook Topics Safeguards*
LV with high tension (HT) outlet	High voltage (HT) outlets	(Y12+)	Use only if students are aware of 'hazardous live' dangers and are under close supervision.	
		N	LV units with HT outlets are best not used unless specifically required. See HT below.	
high tension up to 400 V (HT)	Serious shock hazard	(Y12+)	These units can supply up to 150 mA because some are designed for discharge tubes which require currents higher than 5 mA. Only staff who are trained may use or supervise work with this equipment Shrouded plugs must always be used.	*Handbook Topics*
400 V to 6 kV extra high tension (EHT)	Serious hazard with current greater than 5 mA	(Y12+)	Provided the maximum current is less than 5 mA any electric shocks, although possibly frightening, should be harmless. Supervised by competent staff using a supply limited to 5 mA or less.	*Handbook Topics Safeguards*

Table 16.3   Electrostatic machines and induction coil

Equipment or Procedure	Hazard(s)	Suitability	Guidance / Comments	Reference
Van de Graaff generator and Wimshurst machine	Electric shock (this should not be dangerous unless extra capacitance has been added to these machines)	T	Y7+ pupils may participate in 'charging hair' etc. with care.	
		T	The generation of sparks could cause nearby microcomputers to 'crash' and users should be warned.	
		N	Extra capacitance should not be used because of the increased electric shock risk.	
Induction coil	Electric shock (limited unless the current is over 5 mA)	(Y12+)	Use only if the mean output current is limited to 5 mA. (See 'generation of sparks' above.)	
		N	Very old coils or those from industrial sources with an output current of more than 5 mA. Such coils are not suitable.	

Table 16.4   Practical activities

Equipment or Procedure	Hazard(s)	Suitability	Guidance Comments	Reference
240 V AC mains carbon arc	Severe electric shock, especially when making adjustments to the apparatus	T	Only demonstrate using an approved method with safety screens in place and a projected image.	*Handbook Topics*
	Short-wave UV radiation		Carbon arcs should never be unshielded.	
conductivity of glass	Severe electric shock as above	T	Only demonstrate using an approved method with a regularly tested connecting box.	*Handbook Topics Safeguards*
repairs and testing of mains-powered equipment	Electric shock danger, especially for untrained staff attempting testing or repair or working on a poorly insulated floor	T	A competent teacher or technician undertaking testing or repair must have a colleague nearby and there must be a designated earth-free work area with vinyl/rubber flooring and special power supply.	
	Live fault finding/ adjustment	N	Should not be attempted in school.	
Anodising and electroplating	The LV supply is unlikely to be a hazard but the chemicals used and their products will often be so	(Y9+)	A fume cupboard may be needed.	*Handbook Hazcard*
Body monitoring skin contact electrodes	Procedures designed to minimise skin contact resistance may lead to severe electric shock	(Y11+)	Use the lowest possible voltage: always below 12 V for skin contact. Mains-powered ancillary equipment must be isolated (for example by optical isolators) from the sensing circuits.	*Handbook Topics Safeguards*
Cathode ray oscilloscope (CRO) investigations	None, unless currents and voltages to be measured exceed safe limits; see Power supplies	(Y9+)		
'live' repairs	Serious electric shock if the outer case is removed	N	Live fault finding/adjustment should not be attempted in schools.	

Table 16.4   Practical activities - *continued*

Equipment or Procedure	Hazard(s)	Suitability	Guidance Comments	Reference
Discharge tubes and 'Teltron' tubes	See Power supplies	T		
Electrophoresis	Electric shock risk where voltages exceed 25 V	T	Use apparatus operating at less than 25 V, or designed to prevent connection until solutions are covered.	
Model transformer (power line) demonstration				
uninsulated lines using 240 V AC	Severe electric shock possible between either conductor and earth	N	Alternative method required.	*Handbook Topics*
uninsulated line using DC	Severe electric shock possible if lines connected to high-voltage DC.	N	Alternative method required.	*Handbook Topics*
insulated floating lines with AC transformed to approximately 240 V	None if units are properly constructed	T	Demonstrate using correctly constructed units tested regularly.	*Handbook Safeguards Topics*
power line at 50 V AC	Small electric shock risk	T	Insulation advisable. Ensure that the voltage cannot be increased.	
LV power line maximum 25 V	None, unless the input voltage can be increased	(Y10+)	Provided the input is fixed and pupils follow instructions carefully.	
Transformers – other demountable demonstration	240 V AC mains in primary coil	T	Demonstration should be done by competent staff using coils and connectors designed to minimise risk.	*Handbook*
	Poor insulation or 4 mm connectors on old models	N	Models with 4 mm connectors should not be used with mains supply.	
	Transformation to voltage above 25 V	T	Only to be done if accidental contact is prevented.	
plug-top (mains power adaptor)	If plastic case is damaged, live parts at 240 V AC may be accessible	(Y7+)	Frequently used for computers, games etc. so pupils need to learn safe working practice.	
variable transformer	Electric shock; the output ranges up to 270 V AC.	T	Shrouded 4 mm connectors or mains connectors are essential. Staff should be trained in the safe use of such equipment.	*Handbook*

Table 16.5   Plugs and sockets

Equipment or Procedure	Hazard(s)	Suitability	Guidance Comments	Reference
Domestic electricity meter	Access to exposed live parts in poorly mounted or unmounted models	T	Use only if the meter is permanently mounted to prevent access to live parts. Consider an RCD integral with mains plug.	*Topics*
Extension leads	Trailing cable	T	Trailing extension leads should be avoided except as a temporary measure. Extension cables should be fully unwound from a reel before use.	*Handbook*
	Overheating using connecting cable/flex rated below 13 A or with cable coiled up when in use	T		
Multi-socket outlets (multiblock)	As above Overloading with unfused types	(Y10+)	Should not be used as a substitute for permanent wiring. The maximum load of 13 A should not be exceeded.	
Adaptors	Overheating	N	Use a correctly-fused multi-socket outlet instead of an adaptor.	
Wiring 13 A plugs domestic wiring models	Severe electric shock where a length of flex is connected to a standard plug and that plug is inserted into a live socket	N (Y9+)	This method is too hazardous. It is made safe by: 1. turning off the mains supply to all accessible sockets: 2. using 'off-standard' or disabled plugs which do not fit the sockets; 3. insulating permanently the 'open' end of the flex.	*Handbook* *Topics*
		T	It is useful to demonstrate correctly connecting a plug to an existing piece of equipment. Always use plugs with shrouded pins.	*Handbook*

Table 16.6　　Heating and lighting

Equipment or Procedure	Hazard(s)	Suitability	Guidance Comments	Reference
Electric fire	Electric shock from exposed element, burns and fire	N	Bare element fires should not be allowed in science areas for any reason. Replace with silica-insulated version.	
Fan heater	As above	N	While these may be used as emergency space heaters, they are not suitable for other applications.	
Incubators with heaters	Danger of skin burns	(Y7+)	Choose a type supplied for use in schools.	*Handbook*
Infra-red lamps	Electric shock, burns, fire and risk of explosion if wetted	(Y10+)	Not to be left unattended and particular care needed if used over water.	
LV immersion heaters	Explosion risk caused by ingress of water, especially if allowed to cool while submerged	(Y9+)	Check cable sealant is intact and do not allow submersion of heater top, unless it is of a recent design.	
'Radiant' heaters	Electric shock from exposed spiral element, burns and fire.	N	All 240 V AC radiant heaters with mesh covers, no matter how fine, are unsuitable. Replace with LV version or use a carbon-filament lamp.	*Handbook* *Safeguards*
Soil heaters	Possibility of skin burns	(Y7+)	Provided it is a type supplied for use in schools.	
Aquaria				
aerators	Electric shock especially from handling wet connecting cables, sockets and switches. Siphoning back of aquarium water.	T	DIY installation and home-made equipment not up to professional standard should not be used.	*Handbook*
infra-red lamps	See above			
lighting	As above for aerators, especially if units sited directly above tanks of water	(Y7+)	As above. Fluorescent lights preferred, sealed within hoods so that water is excluded.	

Table 16.6   Heating and lighting - *continued*

Equipment or Procedure	Hazard(s)	Suitability	Guidance Comments	Reference
overnight running	Overheating, fire and breakdown	T	Only use equipment intended for such continuous use, make sure procedures are clear and responsibilities known. Warning notices may be needed to prevent unintentional switching off.	*Topics*
pumps and heaters	Electric shock, especially from handling wet connecting cables, sockets and switches	(Y7+)	Only installations supplied for use in schools should be allowed. Pupils should be supervised when assisting with cleaning, feeding, etc.	*Handbook*
vivaria	As above; also, rapid evaporation may render a thermostat inoperative	(Y7+)	As above.	

Greenhouses

electrical supply	Electric shock		The supply must be via an RCD with splash-proof connectors.	
heaters	Electric shock especially from handling wet connecting cables, sockets and switches while standing on earth, especially when wet or in contact with metal frame	T	Only heaters intended for greenhouses should be used. Be aware of the danger of damage to cables. The alternative of using paraffin or portable gas heaters is not less hazardous.	
lighting	As above	T	DIY installation and home-made equipment not up to professional standard should not be used.	
overnight running	See Aquaria above	T	Particular care may be needed if pupils can gain access to greenhouses outside school hours.	

Table 16.7   Power tools

Equipment or Procedure	Hazard(s)	Suitability	Guidance Comments	Reference
Electric drills (hand held)	Damage to mains lead, mechanical injury, especially to eyes	T	Always wear eye protection and use hand-held power tools for small jobs only.	*Handbook*
Electric saws and sanders (hand held)	As above, plus dust inhalation	T	Always wear eye protection and use hand-held power tools for small jobs only, and limit the amount of dust produced. Guards, as supplied, must be fitted.	*Handbook*
Electric power tools (mounted)	Mechanical injury, especially with circular saws. Clothing, etc. may become entangled	N	Work requiring such equipment is best left to trained workshop personnel.	
Glue guns	Electric shock danger from poor quality construction. Skin burns from lack of heat insulation round the nozzle.	T (Y7+)	Ensure they are constructed to 'double insulation' standard or properly earth-bonded. Check that misuse does not lead to melted cable insulation.	
Soldering irons 240 V AC	Burns, electric shock from poor earth connection or melted insulation.	T	Check for damage to cable. Use LV (especially 12 V AC) type for pupil use and heat-resistant cable.	*Handbook*
	Fumes from the flux in 'cored' solder are a known respiratory sensitiser and may trigger asthma attacks		HSE now recommend that solder with rosin-free flux should be used.	

Table 16.8   Domestic/commercial appliances

Equipment or Procedure	Hazard(s)	Suitability	Guidance Comments	Reference
Dishwashers	Chemical residues in glassware	T	Only a teacher or technician should have access, and glassware should be rinsed before being loaded.	
Hair driers and hot air blowers	Electric shock, especially in contact with water	(Y10+)	It is best to use those supplied for use in schools and others only after inspection and testing.	
	Fire risk	N	Hot-air paint strippers are not suitable. The fire risk with a hot air blower is much greater than with a hair drier.	
Heating and lighting				
Hotplates and ovens	Burns, especially when misused	(Y7+)	Do not allow pupils unsupervised access to hotplates and ovens. Site carefully to avoid casual contact.	
	Mercury vaporised from broken thermometers	T	Spill kit may be needed. Replace with electrical or mechanical thermometer.	*Hazcard Handbook*
Incubators and soil heaters				
Mixers/blenders	Whirling blades; risk of fire or explosion if used for mixing highly flammable liquids, for example hexane.	T	May be used for the preparation of plant and animal materials.	*Safeguards Hazcard*
Outdoor use	Electric shock	T	Any equipment temporarily used outdoors should be supplied through an RCD.	
Overnight running				
Refrigerators with flammables	Risk of fire or explosion if used for storing highly flammable liquids, for example ethoxyethane (ether)	N	The switches, door lights, etc. are not spark-proof in domestic versions.	*Handbook Topics Hazcard*
Vacuum cleaners	Whirling blades and brushes Dust blown into face and eyes	(Y10+)	Cylinder models may be used with air tracks and for Bernoulli effects.	
		N	Upright models should not be used other than for cleaning tasks.	

16.2   References

1   Safety requirements for electrical equipment for measurement, control and laboratory use, IEC 1010-1, 1990 (BS EN 61010-1, 1993) British Standards Institution.

2   Electrical Safety in Schools (Electricity at Work Regulations 1989) Guidance Note GS23, revised 1990. HSE.

3   Electromagnetic Compatibility Regulations 1992.

4   Electrical Testing 1980, HS(G)13. HSE.

5   Provision and Use of Work Equipment Regulations 1992.

6   Medical Aspects of Occupational Asthma, 1991, Guidance Note MS25, HSE.

7   The Electricity at Work Regulations 1989.

8   Memorandum of Guidance on the Electricity at Work Regulations 1989

9   Electricity at Work - Safe Working Practices, HS(G)85.

17 USING LIVING ORGANISMS

The COSHH Regulations 1994 place a duty on employers to make an assessment of risks for work with substances hazardous to health and to take steps to prevent or control adequately the exposure of employees to these substances. Employers must provide information, instruction and training for employees who may be exposed to such substances, in particular, the risks to health and any precautions that should be taken.

17.1   General safety considerations

Very useful and interesting practical work can be carried out with living organisms or material of living origin. This presents little risk provided that the basic principles of good hygiene are followed:

- hand washing is particularly important;
- no material or contaminated fingers should ever be put in the mouth; and
- material which, under, everyday circumstances would be considered edible, for example yoghurt and parts of many plants, must never be consumed unless prepared and presented under conditions in which good hygiene can be ensured.

Conditions of good hygiene are unlikely to be found in science laboratories. At all times adhere to good laboratory practice:

- any activity in which there might be a hazard requires a risk assessment;
- attention should be paid to the purchase of safe resources;
- suitable protective measures should be taken to prevent exposure to infection from micro-organisms either orally, by skin penetration or contact or from splash;
- brief instruction may be needed before certain resources are used with anyone unfamiliar with them; and
- animals maintained in schools must not be, or become, a hazard to either physical or mental health as a result of infection or their use and perceived treatment.

In addition to these principles, other special precautions are necessary for particular procedures or when studying certain living organisms. Some materials of living origin can act as sensitisers; once exposed to them, some individuals will react to much smaller exposures in future. Locusts are a well-known example, and thus particular care is needed when cleaning out their cages.

17.2   Animals

continued on page 125

Table 17.1   Animals: diseases and physical injury

Source of Hazard(s)	Guidance	Reference
Transmissible diseases (zoonoses)	Diseases caused by micro-organisms or parasites may be transmissible between animals and humans. Infection can occur by ingestion, inhalation, through the mucous membranes and skin, especially if this is cut, and as a result of bites. Living and dead animals or their parts may be infected. The surroundings of the animal may also be infective, for example soiled bedding, cages and the water in aquaria.	
Infection from wild animals	Animals known or likely to be infected should not be brought into schools. Such animals include native British wild mammals and birds (alive or dead). Recently imported wild animals may also be infected; examples include psittacine birds (budgerigars, macaws, parakeets and parrots) and the giant (African) land snail (*Achatina fulica*). Animals should not be taken from the wild unless they present a minimal risk to health, for example wood lice. (In addition, the conservation of wild populations must always be considered; there may also be legal constraints on the collection and keeping of animals.) Wild animals thus collected should be returned to their natural habitat as soon as possible.	
Injured wild animals	Injured wild birds, and occasionally other animals including mammals, are often brought into schools by pupils. It is rarely possible to save them and it is best to assume that they are infected. They should be kept isolated and handled using gloves, humanely destroyed and disposed of correctly; (see Disposal). If this is not possible in school, then a local veterinary surgeon or a branch of the PDSA or RSPCA should be contacted.	
Infection from animals bred in captivity	This is unlikely, if animals are obtained from reputable sources from which the quality of the stocks can be assured, for example, the Laboratory Animals Breeders Association (LABA) accreditation scheme for particular mammal species.	*Handbook; Be safe!; Safeguards; LABA; (CLEAPSS has a copy of the LABA scheme register)*
Parasites	No stages in the life cycle of invertebrate species which are parasitic in vertebrates, especially humans, for example, the tapeworm, (*Taenia spp*) must be brought into schools unless specially prepared for display.	
Bites and scratches	Infection can be transmitted through bites and scratches from animals and skin abrasion from infected cages. A record should be kept of all such incidents. If the animal stock has been correctly maintained and hygiene is good, bites and scratches are usually not serious. They should be washed carefully and protected with a non-allergenic adhesive dressing. Medical advice must be sought if there are doubts about the risk of infection.	
Field work	If small mammals are trapped, there is the risk of infection from such animals. Good hygiene is most important and suitable protective gloves should be worn to protect against bites and scratches. In addition, water in canals, rivers and ponds may be infected, for example with *Leptospira spp*. causing Weil's disease.	

Table 17.1   Animals: diseases and physical injury - *continued*

Source of Hazard(s)	Guidance	Reference
Maintenance of animals stocks	Correct management, careful handling and good hygiene are needed to maintain animals in good health and free from disease. At the end of any practical activity with animals, hands should be thoroughly washed with soap and water. The routine of maintenance must be continuous during school holidays or if animals are 'boarded out' during these times.	*Handbook*
Cleaning cages and containers	Cages and containers must be regularly cleaned. The use of a suitable disinfectant is recommended. With aquatic organisms no traces of this should remain. Surfactant disinfectants, for example, BAS/ASAB/Tego are suitable for routine use. In cases of severe contamination and for containers of aquatic organisms, chlorine-based disinfectants may be best (for example, sodium chlorate(I) (hypochlorite) solution at not less than 1% concentration at time of use) but they must be in contact for at least 15 minutes and kept to concentration since degradation occurs particularly when in contact with organic debris. Note that hypochlorite solution attacks metals. Protective gloves and clothing should be worn when cleaning cages and containers.	*Handbook Safeguards Topics*
Electrical heating, illumination or aeration of aquaria, cages or vivaria		
Cross-infection	Wild animals should be prevented from coming into contact with stock, particularly wild rodents with small mammals, and with foodstuffs, bedding and litter materials. These materials should be kept in secure containers and animal stock suitably protected, ideally in an animal room or house. 'Visiting animals', such as domestic pets or animals maintained by a 'visiting animals scheme' should be kept strictly apart from the stock kept in schools to prevent the possibility of cross-infection. If, during school vacations, animals must be taken off the school premises to be looked after, particular care must be taken with small mammals and birds so that they are not infected by wild or domesticated species. Pairs or colonies of vertebrate animals should not be separated as this may lead to later difficulties on re-pairing.	
Materials for dissection	No whole or parts of dead vertebrates (such as hearts, kidneys, livers and lungs) should be brought into schools unless they have been obtained from an abattoir, butcher or fishmonger (and thus subject to a prior inspection by the Local Health Authority), or they have been specially prepared for dissection or for display so that there is no longer a risk of possible infection. In view of the slight theoretical risk of the transmission of bovine spongiform encephalopathy (BSE), bovine eyeballs **must** not be dissected. The eyes of pigs or sheep, are preferred. The possibility of contamination from the dead animal material, of pieces of hard tissue such as cartilage being flicked into pupils' eyes, and cuts from scalpels, must be guarded against.	*Handbook DFEE letter to CEOs Bovine eyeball dissection in schools 25.7.95*

Table 17.1   Animals: diseases and physical injury - *continued*

Source of Hazard(s)	Guidance	Reference
Disposal	Animals must be humanely killed. Dead bodies and other remains together with contaminated material, including soiled bedding and litter materials, must be disposed of so that there are no subsequent health hazards. Disposal by incineration is a preferred method. Where this is not possible and death by disease is suspected, it may be necessary to seek advice from the local council refuse disposal scheme or a vet. 'Sharps', such as scalpel blades used in dissection, should be placed inside a container (such as a tin can) prior to disposal so that there is no risk of cuts being caused when refuse is handled.	*Handbook* *Safeguards* *Topics*

Table 17.2   Animals: Allergies and poisons

Source of Hazard(s)	Guidance	Reference
Allergic reactions	Allergic sensitisation can occur to a wide range of materials including the dust from the skin, hair and feathers of animals, in particular mammals, birds and some insects, and also to sawdust and the dust from hay and straw used as bedding or litter.	*Safeguards*
	There may be a long period between the exposure to an allergen and evidence of the development of sensitisation. Symptoms vary in severity and can appear as dermatitis, asthma or irritation of the membranes of the eyes, ears and nose. Once sensitisation has developed and the allergen responsible identified, further contact with the allergen should be avoided as far as possible.	*Handbook*
Skin, hairs and spines	Some hairy caterpillars, for example those often called 'woolly bears', and in particular those of the browntail moth, *Euproctis chrysorrhea*, have hairs which can penetrate the skin and then break off. These can cause an intense local irritation and possible allergic reaction. If such caterpillars have to be handled, protective gloves should be worn.	
Venoms	Venoms, for example from ants, bees and wasps, can be allergenic in addition to their poisonous effects. Every precaution must be taken to avoid stings. Some people become acutely sensitised to bee venom and show signs of severe shock when stung. Immediate medical attention is required with such cases. Poisonous animals, likely to bite, for example many reptiles, should not be kept in schools.	*Safeguards*

Table 17.3   Animals: possible risks to mental health

Source of Hazard(s)	Guidance	Reference
Emotional disturbance of pupils	Whenever living things or materials of living origin are used, the perception by pupils of this use must be considered. Emotional disturbance as a result of, for example, shock on the sight of a dissected animal or the death of an animal should be anticipated and avoided by appropriate means. This disturbance may also arise from the use of visual material such as photographs, films or videos showing dissection or surgery.	
	Reference must be made to any specific directives on the dissection of animals issued by the employer.	
	There should be no pressure on pupils to dissect or to watch a dissection.	
	The well-being of pupils must be considered paramount.	

17.2.1 Suitable animals A wide variety of animals are suitable for keeping and studying in schools, but only if they can be looked after appropriately. The table below gives some examples of suitable animals. The list is not intended to be exhaustive; animals not included will often be entirely suitable. For certain animals, particular care must be taken; for these, advice is given in the column headed 'Guidance'. Refer to CLEAPSS Handbook for further information.

Table 17.4   Suitable animals

Animal Group	Selected Examples	Guidance
Mammals	Guinea pigs Mice Mongolian gerbils Rabbits Rats Russian & Syrian hamsters	Guard against the development of allergies, particularly in those who handle the mammals regularly. If allergic reactions do occur, take immediate remedial measures.
Birds	Chickens & ducks (for studies of egg incubation and hatching)	There is a very small risk from Salmonella bacteria; normal practices of hygiene will guard against disease transmission. If other birds, such as small finches and waxbills, are to be kept, these may be carriers of the disease ornithosis (psittacosis in psittacine birds such as parrots). Animals which have come into contact with wild birds could be infected.
Reptiles	Garter, corn and rat snakes Leopard geckos	Salmonella bacteria may be carried by reptiles; good hygiene is again required, especially if aquatic reptiles, such as terrapins, are kept.
Amphibians	Axolotls Tadpoles of common species of frogs/toads and also bull frogs Clawed toads Tiger salamanders	Note that bull-frog tadpoles will mature into large adults which require specialist care.
Fish	Many species of cold and tropical water	Good hygiene needed. Take care with aquaria involving electrical equipment for heating, lighting, filtration etc.
Insects and arachnids	Butterflies and moths Crickets Locusts Stick insects Fruit flies Tropical spiders	Beware of lepidoptera species with 'hairy' larvae which may cause allergic reactions if handled. Locusts may also produce a severe allergy; they should not be kept in continuous culture and care should be taken to avoid raising dust when cleaning cages. The two-striped or Florida stick insect may spray an irritant chemical when handled. Tarantula spiders rarely bite but care should be taken when handling the animals and vigilance shown for evidence of allergy developing.
Molluscs	Garden slugs and snails Giant African snails	Avoid recently-imported giant African snails which may be carrying a parasite transmissible to humans; snails bred in this country cannot be infested. Good hygiene is needed after handling all snails.
Other land invertebrates	Earthworms, giant millipedes, wood lice	
Aquatic invertebrates	Water fleas, Hydra, water lice, fresh water shrimps etc. Protozoans: Amoeba, Euglena etc.	Ensure protozoans studied are not species able to parasitise humans; those from educational suppliers should be safe.

Any animal that will not be cared for adequately should not be kept in schools.

17.2.2   **References**

1   The Educational Use of Living Organisms: A Source Book, 1975, Kelly PJ & Wray JD, The English Universities Press (London).

2   The Occupational Zoonoses, 1993, Health & Safety Executive, ISBN 011 886397 5.

3   Zoonoses: Are they a school problem?, 1983, Robinson DF, School Science Review, 64 228 452-461.

4   Science Teachers' Handbook, Secondary, 1993, Hull, Richard Ed/Association for Science Education, Chapter 9, Simon & Schuster Education ISBN 0 7501 0449 X.

5   Animals & Plants in Schools: Legal Aspects. AM 3/90, 1990, DFE. 6 Small Mammals in Schools, 1989, Royal Society for the Prevention of Cruelty to Animals.

6   Small Mammals in Schools, 1989, Royal Society for the prevention of Cruelty to Animals.

7   Animals in Schools, 1986, Royal Society for the Prevention of Cruelty to Animals.

8   Visiting Animals Schemes, 1988, Royal Society for the Prevention of Cruelty to Animals.

9   Letter to all chief education officers, grant-maintained schools, city technology colleges, independent schools, non-maintained special schools and FE colleges, 25 July 1995, Department for Education and Employment.

17.3   **Plants**

Note that under the provisions of the Wildlife and Countryside Act 1981, unless authorised by English Nature or the Countryside Council for Wales, it is illegal to pick, uproot, sell or destroy certain protected wild plants and to uproot any other wild plant unless permission has been given by the landowner.

Table 17.5   Plants: poisoning

Source of Hazard(s)	Guidance	Reference
Poisonous plants	The dangers associated with certain plants, large fungi and seeds are widely recognised. These are avoided if no plant material is eaten but skin contact alone can lead to poisoning with certain species. The whole of a plant or fungus may be poisonous or it may be only certain parts, such as the leaves or seeds, which present a significant hazard. Less obvious examples of this include rhubarb leaves and potato stems/leaves. Sometimes poisonous plants are particularly hazardous because they resemble familiar but safe species, for example the death cap toadstool which looks and smells like an edible mushroom, and laburnum seeds which resemble peas in a pod. For some edible species, the toxins are only inactivated if material is well cooked, for example red kidney beans. A number of wild berries are poisonous to humans but not to birds.	*Topics* *Handbook* *Safeguards*
Poisoning of laboratory animals	Check that all plant material supplied as food is suitable and clean. The death of locusts from dog nematodes on contaminated grass has been recorded. Use only correct seed foodstuffs since commercial seeds for planting often have a pesticide dressing.	*Safeguards* *Topics* *Handbook*
Using pesticides	'Amateurs', which generally covers all teachers except perhaps those specially trained in rural science, may only use pesticides as available from garden centres and high street stores. All such pesticides include full instructions for use which must always be followed. Wear eye protection and, if necessary, gloves. Wash hands thoroughly after use. No hand to mouth operations should be allowed when pesticides are being used.	*Handbook* *Topics*
Pesticide seed dressings	Commercially-available seeds are usually treated with fungicide. Seeds for use in school should be purchased, where available, from health or pet shops because these seeds are intended for consumption. If commercial seeds for planting are used, then adherence to good laboratory practice will avoid difficulty. In addition, it is possible to remove the seed dressing on, for example, broad beans, by washing in running water for 2–3 hours.	*Handbook* *Safeguards* *Topics*
Pesticides on plants collected from the wild	The dangers of collecting plants treated or contaminated with pesticide are higher at certain times of the year. However, most modern pesticides are short-lived which reduces the risks. It is important to be familiar with the collecting location in order to avoid the time of spraying. In addition, adhering to basic principles of hygiene will minimise actual risk.	

Table 17.6   Plants: allergies

Source of Hazard(s)	Guidance	Reference
Allergic reactions	A number of plants produce chemicals in their sap which cause an allergic reaction in some people. There may be a long time between the exposure to an allergen and the development of sensitisation. The sap from giant hogweed (and some other umbellifers) is known to cause sensitisation of the skin after contact and subsequent exposure to light; the effects are like sunburn. The sap from some bulbs, especially hyacinth, released on handling can produce a skin rash in sensitive individuals. In all cases, the effects can be eliminated by avoiding all species known to present a hazard or, where this is impossible, minimised by wearing protective gloves. If children are known, or believed, to be susceptible, then such precautions are essential. Stinging by nettles is a common example of a reaction to an allergen, so well known that children will naturally avoid these plants. When doing field studies, teachers should avoid sending children into areas of high nettle density. Hay fever is also well known, especially to sufferers. During the hay fever season it is impossible to avoid pollen altogether but teachers should be alert to the reactions of sufferers when carrying out field work or when studying grasses in the laboratory.	*Safeguards* *Handbook*
Fungal spores	Most fungi produce a vast number of spores which can produce allergic reactions in some individuals when inhaled or ingested. When studying larger fungi such as mushrooms, risks can be minimised by limited handling and by choosing specimens which are not at the spore-releasing stage. Fungal colonies should be studied in closed containers which can then be disposed of unopened or after sterilisation.	*Topics* *Safeguards* *Handbook* *Micro*
Transferring allergens and disease on fingers to mouths and eyes	In exercises involving handling of plant material remind pupils of the dangers and strictly enforce hygiene practices. In some areas, for example on the banks of streams and lakes, plants may have been contaminated by bacteria from wild animals.	*Handbook* *Safeguards*

Table 17.7　　Plants: investigations with plants involving chemicals

Source of Hazard(s)	Guidance	Reference
Extraction of chlorophyll from leaves	The extraction of chlorophyll with hot alcohol (ethanol, methylated spirits) offers a significant risk of fire. Never use ethanol in the presence of naked flames; always heat it by using a hot water bath preferably filled from a kettle and certainly with any Bunsen flame extinguished. The extraction of chlorophyll for chromatography using other solvents (for example, propanone) may present even more of a fire hazard.	*Safeguards*
Using growth substances	Pure indoleacetic acid is likely to be toxic since closely related compounds certainly are. It should be handled using gloves and eye protection. When diluted for pupil use, concentrations are low enough to present no risk. 2,4-dichlorophenoxyacetic acid (2,4-D) is toxic and must be handled with great care following all instructions carefully. Avoid skin contact.	*Handbook* *Safeguards*
Studying chromosomes	Colchicine, used in the past for arresting cell division, is now considered too dangerous for use in schools. An alternative, which is itself classified as harmful and irritant, is 1,4-dichlorobenzene (p-dichlorobenzene, Paradichlor).	*Handbook* *Safeguards*
Studying plant tissues	A number of solutions are used for macerating plant tissues prior to further examination. They are generally classified as harmful, corrosive or flammable and should be handled with appropriate care and, where necessary, using protective equipment. Instructions for their use must be carefully followed. This advice also holds for clearing fluids, fixatives, mountants, preservatives and stains.	*Handbook*

Table 17.8   Plants: physical injuries from using or growing

Source of Hazard(s)	Guidance	Reference
Spines and thorns	Cacti are the most obvious sources of spines. Injuries often happen when handling specimens which appear to have soft spines when in fact this downy surface masks an abundance of smaller spines. All cacti need very careful handling by wrapping in a layer of protective material. Limit the amount of general handling of such plants and be particularly careful with younger pupils and those for whom basic discipline cannot be guaranteed. Only bring other thorny plants, for example brambles or parts of such plants into the laboratory for particular activities, keep them securely and dispose of them as soon as possible. In the event of a pupil receiving an injury, remove obvious spines with a clean pair of forceps or tweezers and make sure the affected area is washed carefully with soap and water. If it is not easy to remove all the spines, then medical advice must be sought.	*Handbook* *Safeguards*
Cutting sections of plant material	This requires a sharp razor blade or scalpel. Before starting this activity, it is important to consider the skill levels and discipline of the pupils. If either cannot be guaranteed then the use of pre-cut sections would be prudent.	
Cuts from glass	Greenhouses, cold frames and other glass propagators present risks from broken glass. Children should not be allowed to play in the vicinity of such structures nor, in general, to handle broken panes of glass. However, older children can be taught how to replace broken panes, if gloves and eye protection are worn.	
Electric shocks	Growing plants often involves electric lights or electrical heat sources, for example aquaria, greenhouses or bench lamps. All electrical fittings must be correctly wired and pupils must not be allowed to interfere with or repair them.	

Table 17.9   Plants: some hazardous forms

Plant	Hazard	Guidance	Reference
Buttercup *Ranunculus spp*	Poisonous	All parts	*Handbook*
Castor oil plant *Ricinus communis*	Poisonous	Seeds are highly poisonous	*Safeguards*
Chilli peppers *Capsicum annum*	Allergenic	Seeds and fruits	*Safeguards*
Chrysanthemum *Chrysanthemum spp*	Allergenic	Sap	*Safeguards*
Deadly nightshade *Atropa bella-donna*	Poisonous	All parts	*Safeguards*
Deathcap toadstool *Amanita phalloides*	Poisonous	All parts	*Handbook*
Foxglove *Digitalis purpurea*	Poisonous	All parts	*Handbook*
Grasses *Graminae*	Allergenic	Pollen causes hayfever	
Giant hogweed *Heracleum mantegazzianum* and other umbellifers	Allergenic	Sap causes photosensitisation	*Handbook*
Hyacinth *Hyancinthus orientalis*	Allergenic	Sap from bulbs	*Handbook*
Laburnum (golden rain) *Laburnum anagyroides*	Poisonous	All parts but especially seeds	*Safeguards*
Lilies and Liliacea family	Allergenic	Sap	*Safeguards*
Peanut *Arachis hypogea*	Allergenic	Seeds, oil and peanut butter can lead to amaphylactic shock in extreme cases.	
Potato *Solanum tuberosum*	Poisonous	Stems and leaves (and green tubers)	*Safeguards*
Primula *Primula spp*	Allergenic	Sap	*Safeguards*
Red kidney bean *Phaseolus vulgaris*	Poisonous	Beans, unless well cooked	
Rhubarb *Rheum rhaponticum*	Poisonous	Leaves	*Safeguards*

17.3.1 References

1 Animals and plants in schools: Legal Aspects. AM 3/90, DFE.

2 Poisonous Plants and Fungi, 1988, Cooper MR & Johnson AW, Ministry of Agriculture, Fisheries and Foods, HMSO, ISBN 0 11 242718 9.

17.4 Microbiology and biotechnology

17.4.1 General safety considerations

Work in microbiology and biotechnology in schools is categorised into three levels which are described in outline below. Although appropriate for use in schools, these levels are not the same as 'levels of containment' used by professional microbiologists. Further detailed guidance for each is provided below.

- Level 1 (L1): work with organisms which have little, if any, known risk and which can be carried out by teachers with no specialist training. The organisms will be observed in the closed containers in which they were grown.

- Level 2 (L2): work where there may be some risks of growing harmful microbes but these are minimised by careful selection of organisms or sources of organisms and by culturing in closedcontainers which are taped before examination and remain unopened unless the cultures within have been killed. Once a culture prepared by pupils has been grown, subculturing or transfer of organisms from one medium to another may not be done. L2 work may be carried out with pupils between the ages of 11 and 16 years (KS 3 and 4) and by science teachers who may require training and some supervisio n. This can be provided through a short in-service course or in school by a knowledge able biology teacher.

- Level 3 (L3): work where cultures of known fungi and bacteria are regularly subcultured or transferred. This work is normally confined to students over the age of 16 and institutions where facilities are appropriate. Teachers should be thoroughly trained and skilled in good aseptic techniques. This is a higher level of training than required for L2 work. Non-specialist teachers should not carry out or supervise this work.

A significant risk associated with work in microbiology or biotechnology is the generation of microbial aerosols, where fine droplets of water containing cells and/or spores of microbes are released into the air. Aerosols can be formed whenever liquid surfaces are broken or material is crushed or ground. The particles are so small that they are easily carried by air currents and can penetrate into the respiratory system. Many of the safety measures detailed below are designed to minimise the risk of aerosol formation.

Although microbiology and biotechnology are considered separately in the following sections, a major difference is one of scale with the corresponding increased risk of generating large amounts of unwanted micro-organisms. Some additional hazards for biotechnology are also described.

Before work with microbes is started, pupils should wash their hands with soap and water (except for L2 and L3 work investigating microbes on unwashed hands) and cover any cuts with waterproof plasters. Hands should also always be washed after working with microbes.

Table 17.10 Microbiology

Source of Hazard(s)	Guidance	Reference
Organisms	L1 Limited to yeasts used for baking and brewing, algae and some moulds and commonly-occurring bacteria where they grow naturally on decaying food or plant material.	*Be safe! Safeguards*
	L2 Known organisms including selected bacteria and fungi from recognised suppliers. Organisms may be cultured from the environment but not from environments which are likely to contain harmful organisms, for example lavatory seats or body surfaces other than fingers or hands. Containers of such cultures must be sealed before examination (but only after incubation).	*Topics Micro Handbook*
	L3 Known bacteria and fungi from a reputable source. Organisms may be cultured from the environment or from body surfaces providing the work is appropriate to the course and that cultures are not opened by students.	
Culture media	L1 Organisms can only be cultured on the substances on which they grow naturally, for example bread, fruit, vegetables, milk, cheese, yoghurt, hay or grass and other plants, in the case of rusts and mildews.	*Be safe!* *Micro Topics Handbook*
	L2 Agar-based culture media generally with a simple nutrient base, low pH or high salinity, but not those which select for organisms which are potentially pathogenic to humans, for example blood agar, McConkey's agar, dung or faecal agar. Similar restrictions apply to nutrient broth solutions.	
	L3 As for L2.	
Storage of organisms and media	It is unwise to store organisms in schools for any length of time, except perhaps to maintain cultures for some work at L3. Such organisms should be subcultured every 3 months or so but only if completely aseptic techniques can be guaranteed. Organisms should be stored in a refrigerator but never in one used to keep human foodstuffs. Media should be stored as dry powder or tablets. Once sterilised, media can be stored for several months in tightly-sealed, screw-topped bottles.	*Micro Topics Safeguards*
Contamination of teachers, pupils and students	Before beginning practical work hands should be washed with soap and warm water, and all should be washed again after practical work. There must be no hand-to-mouth operations such as chewing, sucking, licking labels or mouth pipetting.	
	L3 Teachers, technicians and students should wear lab coats or aprons which can be relatively easily disinfected and cleaned. Teachers should consider the use of lab coats or aprons for L2 work.	
Inoculation of cultures	Inoculation should involve precautions to prevent contamination of the person and work surfaces. It should also avoid the contamination of culture media with unwanted microbes. Media and Petri dishes etc. should be pre-sterilised or sterilised before use.	*Micro Handbook Topics*
	Media must not be inoculated with material likely to contain sources of human pathogens.	*Micro Topics*
	L3 For transfer of cultures, work surfaces should be swabbed with a suitable disinfectant before and after all operations. Arrangements should be made to sterilise inoculating loops and spreaders before and after inoculation, and to provide discard pots for pipettes and syringes. The mouths of all containers, tubes, flasks, McCartney bottles etc., should be flamed after removing caps and before their replacement. Lids of Petri dishes should be opened only just enough to allow the inoculating tool to be manipulated and for as little time as possible.	*Handbook Topics*

Table 17.10   Microbiology - *continued*

Source of Hazard(s)		Guidance	Reference
Bench surfaces		Benches should be wiped down with a cloth soaked in a suitable disinfectant after all practical work *and* sufficient time allowed for disinfection to occur.	*Micro Topics Handbook*
Incubation	L1	Incubation should be limited to normal conditions in the classroom. The only exception will be yoghurt making at 43 °C, which, by using a starter culture and restricted medium is less likely to encourage unwanted pathogenic growth. Yeast cultures generate considerable quantities of carbon dioxide gas. Incubation containers should be plugged with cotton wool which will allow excess gas to escape.	*Be safe! Micro*
	L2 L3	The upper limit for general school-based work should be 30 °C to avoid the selection of organisms which are adapted to human body temperature. Exceptions to this will include yoghurt making (incubation temperature 43 °C) and the culturing of *Bacillus stearothermophilus* which requires an incubation temperature of 65 °C, too high for human pathogens. Petri dishes should be incubated inverted to avoid condensation dripping onto cultures. During incubation, the lid of the Petri dish should be taped to the base with two or four small pieces of tape so that the lid cannot be accidentally removed and conditions inside cannot become anaerobic. Unless cultures are known to be minimum risk, teachers should consider sealing incubated dishes with tape before examination by pupils or students.	*Micro Topics Handbook*
Spills		All spills should be reported to and dealt with by the teacher, who should record all incidents. All spills carry a risk of aerosol formation and dealing with them must reduce this as far as possible. Spills should be covered with towels or a cloth soaked in a suitable freshly-prepared disinfectant, preferably a clear phenolic or alternatively, freshly-prepared 1% sodium chlorate(I) (hypochlorite) solution, and left for at least 15 minutes. The spill debris should then be swept up using disposable paper towels. Disposable plastic gloves should be worn. Seriously contaminated clothing should be disinfected before laundering. Contaminated skin should be carefully washed with soap and hot water. Lysol and other cresolic disinfectants are caustic and should not be used. Uncapped cultures should not be centrifuged.	*Topics Micro Handbook*
Observation of cultures	L1	Cultures should be viewed in the unopened containers in which they were grown.	*Handbook*
	L2	Cultures should be examined in closed containers which are taped or sealed before pupils examine them. If it is necessary for pupils to open cultures for examination, the latter must first be killed by the teacher or technician placing a filter paper moistened with 40% methanal (formalin) solution in the dish, in the inverted position, 24 hours before examination (care; eye protection and gloves are necessary).	
	L3	Cultures of known and safe microbes can be examined using a variety of techniques. Organisms cultured from body surfaces or any environmental source must be examined in unopened containers, or killed before examination as described above.	

Table 17.11   Biotechnology

Source of Hazard(s)		Guidance	Reference
Organisms		The level restrictions which apply for microbiology also apply to biotechnology work.	*Be safe!* *Topics* *Micro* *Handbook*
	L1	Suitable organisms include yoghurt bacteria, yeast such as for wine or bread and some unicellular algae.	
	L2	Other than L1 organisms, it is recommended that organisms with unusual growth requirements are used, for example high salt, low pH, low temperature. Some examples of minimum risk organisms include *Vibrio natriegens (Beneckea natriegens), Photobacterium phosphoreum* and *Acetobacter aceti.*	
		Teachers wishing to use organisms not listed as minimum risk must have had suitable training in microbiological techniques and should consult an appropriate advisory body.	
		Avoid the large-scale culture of organisms which produce antibiotics, particularly penicillin.	
		Cultures of organisms should only be obtained from recognised suppliers, including culture collections.	
	L3	As for L2.	
Media		The solutions generally used in biotechnology work present few problems other than those associated with quantity and the potential for contamination. Keep quantities to a minimum to make handling easier and reduce the quantities of enzymes, antibiotics etc. which can be generated. All media should be sterilised prior to use. The use of animal dung for investigations of bio-gas generation is not recommended; use grass clippings inoculated with well-rotted garden compost or pond mud.	*Topics*
Incubation/ fermentation		To avoid the growth of pathogenic organisms, incubation should be as far from 37 °C as possible. For the majority of organisms this will mean below 30 °C but for yoghurt making, 43 °C may be used provided hygienic preparation is followed.	*Topics* *Handbook*
	L2,3	Use of fermenters is limited to these levels.	
		The generation of large volumes of gas (carbon dioxide or methane) is a risk associated with many fermentations. Vessels must be suitably vented to allow the gas to escape but prevent aerosol formation or the entry of unwanted organisms. In the case of methane, the gas must be kept away from naked flames and electrical equipment which can cause sparks.	
	N	Other than work with yeasts, anaerobic fermentations should not be used in schools.	
Contamination		Cultures should be started by inoculation with a significant volume of actively-growing inoculum (for example 20% of total volume).	*Topics* *Handbook*
		All equipment and materials (other than the inoculum) should be sterilised prior to use.	
Spills		Routines for dealing with spills are the same as for microbiology. With fermenters there is the risk of spills of large amounts of liquid culture. All possible steps should be taken to guard against this, for example by using equipment within a spills tray. In the case of gross spills, unless the organism is known to be safe, the lab should be cleared before attempting to deal with the spill.	*Micro* *Handbook* *Topics*
Electrical hazards		Keep all electrical leads, especially mains leads, tidy and site electrical equipment so as to minimise the risk of water getting in.	*Topics* *Handbook*

Table 17.10   Microbiology - *continued*

Source of Hazard(s)	Guidance	Reference
Sterilisation and disposal	All cultures must be sterilised before disposal. This is best done using a pressure cooker or autoclave, if possible in conjunction with autoclavable bags. The caps of all screw-topped bottles must be loosened before cultures and media are sterilised. It is very important that instructions for use of the autoclave are followed in order to maintain sufficiently high temperatures for a sufficiently long time. Pressure cookers are unlikely to be equipped with appropriate instructions for sterilisation and advice should be sought. Teachers and technicians should be trained to follow safe working practices. Seals and safety valves should be checked before each use. Heating autoclaves or pressure cookers with Bunsen burners is not recommended. Rapid cooling and the release of steam to lower the internal pressure quickly to atmospheric pressure may shatter glassware and/or cause liquid media to boil over. Chemical sterilisation is much less satisfactory for the disposal of used agar plates and cultures but can be achieved if a freshly-made clear phenolic disinfectant is used. Cultures and equipment must be opened under the surface of the solution and left for up to 24 hours. Again it is essential to follow disinfectant instructions carefully. Chlorate(I) solution is inactivated by large amounts of organic matter, although if a culture may contain viruses this is the preferred disinfectant, since clear phenolics are only effective against bacteria and fungi. Microwave ovens are **not** suitable for sterilisation of most items (though they are useful to liquify prepared agar media.) After sterilisation, solid cultures can be disposed of, in tied autoclave bags or similar, through the refuse system. Liquid cultures can be flushed away down the lavatory or the sink with lots of water. Culture material should not be allowed to accumulate in open or closed waste traps. Incineration is an acceptable alternative to autoclaving. Note, however, that polystyrene Petri dishes will generate harmful fumes when incinerated; a purpose-built incinerator with a tall flue should be used. Clean glass equipment can be sterilised by dry heat in an oven (160 °C for 2 hours) or, in the case of wire loops, by heating to red heat in a Bunsen flame.	*Be safe!* *Topics* *Micro Handbook*
Radiation	Use to induce mutations in yeasts only; ensure that eyes are protected from UV radiation.	*Topics*

Table 17.11   Biotechnology - *continued*

Source of Hazard(s)	Guidance	Reference
Disposal	All cultures should be sterilised before disposal, preferably in an autoclave. If a fermenter cannot be sterilised complete, add clear phenolic disinfectant, at recommended dilution, to the culture before pouring the contents into containers which can be autoclaved.	*Handbook* *Topics*
Enzymes	Handle all enzymes, whether solid, liquid or cultures which may produce them, with due care. Problems with enzymes increase with quantity as well as variety. Minimise skin contact and use eye protection and disposable gloves for solid or concentrated solutions of lipolytic and proteolytic enzymes. Avoid causing powders to get into the air.	*Handbook* *Topics*
Plant growth substances	Often, wrongly, called plant hormones; many are toxic and some may be carcinogenic. Teachers and technicians should handle solids or concentrated solutions with appropriate care. The very low concentrations used in solutions by pupils present no significant risk.	*Topics* *Handbook*
Animal cell culture N	Work with animal tissue culture is not recommended for use in schools. If teachers wish to do such work they should ensure that cultures are obtained from specialist school suppliers.	*Topics*
Practical work with DNA	Practical work with DNA ('genetic engineering') or, more accurately, genetic manipulation is at present restricted in schools to work which does not result in the formation of new nucleic acid molecules. Acceptable work includes plasmid transfer, plant tumour induction, transfer of natural antibiotic resistance between bacterial strains, the induction of mutations in yeast using UV radiation and enzyme fragmentation of DNA with subsequent separation of components by electrophoresis. The great majority of hazards associated with all such work are with well-known and documented chemicals and procedures, including those which arise from the construction of DIY equipment, such as electrophoresis chambers. Risks associated with the DNA itself, or other nuclear material, are considered to be negligible as long as sources of DNA are restricted to those suggested in genetic manipulation kits or accompanying literature (but this does not apply to kits imported from abroad, some of which include substances or procedures which are not considered safe or appropriate). This advice supersedes that given in *Microbiology: an HMI Guide for Schools and Further Education*.	*Handbook* *Topics*

17.4.2 Suitable and unsuitable micro-organisms

The table below lists selected micro-organisms drawn from recent science teaching projects which present minimum risk given good practice. It can also be found in the CLEAPSS Laboratory Handbook and supersedes the information in Microbiology:an HMI guide for schools and further education and Topics in Safety. It is not a definitive list; other organisms may be used if competent advice is taken. It should be noted that strains of species of fungi can differ physiologically and therefore may not give expected results.

Table 17.12 Selected micro-organisms

Bacteria	Fungi
Acetobacter aceti	Agaricus bisporus
Agrobacterium tumefaciens	Armillaria mellea
Bacillus subtilis+	Botrytis cenerea
Chromobacterium lividum*	Botrytis fabae
Chromatium species	Chaetomium globosum
Erwinia carotovora	Coprinus lagopus
(= E. atroseptica)	Fusarium solani
Escherichia coli+	Fusarium oxysporum
Lactobacillus species	Helminthosporium avenae
Micrococcus luteus	Mucor hiemalis
(= sarcina lutea)	Mucor mucedo
Photobacterium phosphoreum	Myrothecium verucaria
Pseudomonas fluorescens**	Penicillium roqueforti
Rhizobium leguminosarum	Phycomyces blakesleanus
Rhodopseudomonas palustris	Physalospora obtusata
Spirillum serpens	Phytophthora infestans
Staphylococcus albus	Pythium debaryanum
(epidermidis)**	Rhizopus sexualis
Streptococcus lactis	Rhizopus stolonifer
Streptomyces griseus	Ryytisma acerinum
Vibrio natriegens	

(= beneckea natriegens)

Viruses

Saccharomyces cerevisiae	Saccharomyces ellipsoides
Bacteriophage	Saprolegnia litoralis
(T type) (host E. coli)	Schizosaccharomyces pombe
Cucumber Mosaic Virus	Sclerotinia fructigena
Potato Virus X	Sordaria fimicola
Potato Virus Y	Sporobolomyces species

(not the virulent strain)

Tobacco Mosaic Virus

Turnip Mosaic Virus

- This species replaces chromobacterium violaceum and serratia marcescens.
- Some strains have been associated with health hazards. Reputable suppliers should ensure that acceptable strains are provided.
- These organisms have been known to infect debilitated individuals and those taking immunosuppressive drugs.

Alga, protozoa, lichens and slime moulds

Although some protozoa are known to be pathogenic, the species quoted for experimental work

in recent science projects, together with species of algae, lichens and slime moulds, are

acceptable for use in schools.

Unsuitable micro-organisms

A number of micro-organisms have in the past been suggested for use in schools but are no longer considered suitable.

Bacteria	**Fungi**
Chromobacterium violaceum	Aspergillus nidulans
Clostridium perfringens	Aspergillus niger
(welchii) Penicillium chrysogenum	
Pseudomonas aeruginosa	Penicillium notatum
Pseudomonas solanacearum	
Pseudomonas tabacci	

Serratia marcescens

Staphylococcus aureus

Xanthomonas phaseoli

17.4.3	References

For addresses, see Useful Addresses.

1   Organisations which can be consulted about suitability of microorganisms:

Microbiology in Schools Advisory Committee (MISAC)

National Centre for Biotechnology Education (NCBE)

CLEAPSS

ASE

2   Genetic Manipulation Health and Safety Commission Regulations and Guidance, 1978. Health and Safety Executive, HMSO ISBN 0 11 883202 6.

17.5 Using pupils as subjects of investigation

17.5.1 General safety considerations

Pressure must never be placed on pupils to take part in any investigation on themselves. The physical and mental health of individual pupils should be ensured. Hazards are of two types: to mental health, involving emotional disturbance and mental stress, and to physical health. Whenever pupils are involved as the subjects of investigations, the distinct, yet often overlapping, nature of these hazards must be continually borne in mind. Teachers should be aware of the distress that may be caused to pupils when interpretations are made from observations and measurements.

Table 17.13   Hazards to mental health

Source of Hazard(s)	Guidance	Reference
Interpretation of observations and measurements made	Comparison of mental or physical capabilities or size between individuals, or comparisons with so-called 'normal values', or the discovery of a disability such as colour blindness can be distressing. Impressions of fingerprints that are taken should be given to pupils, or destroyed immediately after use.	*Handbook*
Identification of hereditary relationships	This can be distressing, especially if previously unknown.	

Table 17.14   Hazards to physical health

Source of Hazard(s)	Guidance	Reference
Physical exercise which can produce stress	No pupil who is medically excused normal school PE activities should take part in investigations of the effect of exercise on the rate of breathing or pulse rate. However, exclusions should be tactfully organised to prevent any mental stress to pupils.	*Safeguards Handbook*
	Care should always be taken to avoid situations in which competition between members of a group of pupils, for example, during physical exercise or the forced intake and holding of breath, may overstress individuals.	
	Exercise bicycles should be checked before use. Bicycles on rollers are not suitable for exercise testing. Slipping and tripping during step-up exercises should be guarded against.	
Temperature regulation	Investigations on temperature regulation involving the cooling of a part of the body, for example the arm, must always be conducted with great care. Avoid the use of very cold conditions and prolonged exposure.	
Using spirometers and sphygmomanometers	Training is necessary for the safe use of apparatus such as a spirometer or that used to measure blood pressure. It must be emphasised to pupils that any measurements made are approximate only and must not be taken as medically relevant.	*Handbook*
Smelling and tasting chemicals and foods	Pupils should be taught how to smell chemicals safely by gently inhaling very small quantities.	*Safeguards Handbook*
	No substance or food should be tasted unless it is known to be safe and free from possible contamination. Foods or other substances used for tasting must be kept and used under conditions where contamination cannot take place. Investigations must be carried out in such a way that good hygiene is guaranteed and cross-infection between pupils prevented. Foods known to have an irritant or allergenic effect, such as chilli peppers or peanuts, should never be tasted.	
Mouthpieces and thermometers	Mouthpieces, for example, of spirometers, and thermometers used for taking body temperature must be disinfected before and after use using ethanol or a fresh solution of 'Milton' disinfectant at the appropriate dilution, followed by adequate rinsing in water. Alternatively, disposable mouthpieces can be used. They must be immediately disposed of after use.	*Safeguards Handbook*
	Liquid crystal thermometers are available which are placed in direct contact with the skin.	
Saliva	There is little hazard from saliva, and even then only if pupils, teachers or technicians come into direct contact with saliva other than their own. Pupils should wash their own glassware and technicians should wear gloves when handling this. Disinfection of contaminated glassware after investigations (should this be required) can be achieved using a freshly-prepared 1% aqueous solution of sodium chlorate(I).	*Handbook*
Urine	Urine can be hazardous only if pupils, teachers or technicians come into direct contact with urine other than their own, although some may consider it an unsuitable body fluid to use. Pupils should wash their own glassware and technicians should wear gloves when handling this. Disinfection of used glassware using a freshly-prepared 1% aqueous solution of sodium chlorate(I) is recommended.	

Table 17.14   Hazards to physical health - *continued*

Source of Hazard(s)	Guidance	Reference
Cheek-cell samples	Some education employers permit samples of cheek cells to be taken. If so, it is essential to follow a safe procedure, which further reduces the already minute risk of infection, and to make sure that pupils are sufficiently reliable to follow instructions. Samples should be taken using a cotton bud from a newly-opened pack. Disposal of the used cotton bud, slide and coverslip should take place immediately after use in a freshly-made 1% sodium chlorate(I) aqueous solution.	*Handbook*
Blood samples	Blood samples must not be taken from staff or pupils unless permitted by the employer's guidelines. Unless specifically required, for example in some post-16 vocational courses, most education employers do not permit blood samples to be taken. If, exceptionally, it is permitted by the employers' guidelines, sterile conditions must be strictly enforced and all materials contaminated with blood should be autoclaved before disposal or treatment. If samples from a blood bank are investigated, autoclaving before disposal is again recommended.	
Pulsed stimulation	The community health physician and employer must give their approval for any procedure and apparatus used to stimulate muscular contraction artificially. Unpleasant physiological reactions can be induced by rhythmical impulses of light or sound, especially with a frequency of 7–15 Hz. Investigations into this effect must never be attempted.	
Biological feedback	Investigations of biological feedback, in which body rhythms, for example heartbeat or the electrical activity of the brain, are picked up, amplified and then used to influence existing rhythms, must never be attempted.	
Electric shock	Attempts to monitor electrical activity in the body (for example ECG recordings) require the use of low-resistance skin contacts which make electric shocks more serious. Any equipment connected to the body must be very well isolated from hazardous voltages.	

18 USING PHYSICS EQUIPMENT

18.1 General safety considerations

Any activity where there might be a hazard requires a risk assessment. Attention should be paid to the purchase of safe equipment. Brief instruction may be needed before certain equipment is used by anyone unfamiliar with it. One area subject to national regulation is that of ionising radiation, covered by AM 1/92 (see Useful Publications) which every teacher who handles radioactive sources in school must be aware of.

Table 18.1   Mechanical hazards

Equipment or Procedure	Hazard(s)	Suitability	Guidance Comments	Reference
Air guns and projectiles				
air pistols air rifles	Mishandling leading to serious shooting injury especially to eye and head	T	The teacher should be familiar with the demonstration; all present should wear eye protection; guns should be clamped in position.	*Handbook Safeguards*
	Ricocheting pellets		A backstop should be used to catch pellets.	
	Misuse following theft		Guns must be locked away securely when not in use.	
carbon dioxide rockets	Impact injury, especially if no guide wire is used	T	The small carbon dioxide capsule should always be attached to a wire guide (or trolley) before being pierced.	
chemical rockets	Explosion and impact injury	T	Generally travel faster and higher than water rockets but are expensive to run. It is essential to follow manufacturer's instructions.	
water rockets	Pre-launch explosion and impact injury	(Y9+)	Do not modify kits or instructions in any way. Those made from hard plastic frequently shatter on impact.	
Heavy masses				
suspending	Hands and feet directly beneath suspended masses	(Y7+)	Cardboard box full of waste material directly below the suspended mass protects floors and feet.	
handling	Strain or injury caused by lifting	T	Must avoid risk to themselves and to students.	
whirling masses	Injury to bystanders and equipment damage	(Y10+)	Perform outdoors if possible, use rubber or soft-plastic masses.	
Suspension beams, hoists, pulleys and gears	Structural collapse, injury to those below.	(Y10+)	Model pulley systems must be limited to a few kilograms.	
	Trapped fingers and clothing.	T	Regular safety checks if large loads are lifted or suspended.	
		(Y10+)	A bicycle, sometimes inverted, is often used for mechanical advantage and velocity ratio measurements. A safe working area is essential. No attempt should be made to lift a person off the ground with pulley systems without a proper assessment of the risks.	
Trolley runways	Long runways may fall over if badly stored or stacked Handling injuries if heavy	T	Staff should ensure that stacking and carrying is done with minimum risk to themselves and students especially in relation to falling and tripping.	
Wires wires, strings and plastic filaments under tension	Breakage producing whiplash injuries, especially to eyes See pulleys and hoists	(Y7+)	Insist on eye protection as a means of instilling good practice.	*Handbook Safeguards*

Table 18.2   Heating, ignition and explosion

Equipment or Procedure	Hazard(s)	Suitability	Guidance Comments	Reference
Electric kettle for heat capacity	Electric shock, steam scalds and mercury from broken thermometers	T	Provides a simple, convenient way of estimating specific and latent heat capacities.	
Lead shot for heat capacity	Lead dust	T	Avoid inhaling and handling lead dust.	*Hazcard*
Explosive gases mixture demonstration	Violent explosion, shattering equipment	T	Use small-scale polypropylene syringe demonstrations with oxygen supplied via a non-return valve.	
Fuel calorimetry				
Heat treatment of metals	Burns from hot metal Sudden fracture of brittle specimens	(Y12+)	Wear eye protection and use pliers or vice for snap testing.	

Table 18.3   Radiation hazards (including sound)

Equipment or Procedure	Hazard(s)	Suitability	Guidance Comments	Reference
Discharge tubes operating above 5 kV	X-rays	X	Production of X-rays in this way without approval is prohibited.	
		N	Large induction coils should not be connected to discharge tubes.	*Safeguards*
Electric arcs	Ultra-violet radiation	T	If demonstrated (for example carbon arc) shield with thick glass (6 mm or more). Electric arcs should always be shielded from direct view.	
Ionising radiations	Ionising radiation dangers and the leakage of radioactive material	T	Must be trained to the required level of work undertaken.	
		(Y12+)	Pupils under the age of 16 may only watch demonstrations and must not handle sources except rocks and the very weak cloud chamber sources.	
		X	Any procedure not recognised as 'standard' including generating X-rays is prohibited without further training.	
disposal		T	Sealed sources must be disposed of through the National Disposal Service or the supplier: both will charge. Up to 100 g per day of compounds of radioactive elements, such as uranium or thorium, can be made into solution and flushed down the drain with plenty of water.	
Lasers	High-power laser beams entering the eye	(Y7+)	Staff should be aware of the regulations relating to Class 2 and Class 3 lasers. Goggles are not required when Class 2 lasers are used. Advice previously given in DES *AM 7/70* is now obsolete in this respect.	*Handbook Safeguards*
		N	Class 4 lasers should never be used in schools.	*Topics*
Microwaves	Microwave transmitters may pose a hazard to those fitted with heart pacemakers.	(Y11+)	Warning notices may be displayed to indicate the danger zone if in any doubt. Be especially cautious on open days and the like.	
Rhythmic impulses signal generator (pulsed stimulation)	Unforeseen physiological effects; possible epileptic, migraine trigger	T	Special danger is thought to occur between 7 Hz and 15 Hz.	*Safeguards*
stroboscope (light)	'Freezing' of mechanical movement, especially rotation	T	Take care that all are aware that spinning, etc. has not ceased.	
	As above	N	Sensory stimulation should not be attempted.	

Table 18.3   Radiation hazards (including sound) - *continued*

Equipment or Procedure	Hazard(s)	Suitability	Guidance Comments	Reference
Solar energy				
focusing the sun's rays	Risk of fire	(Y7+)	It is instructive to do this outdoors, with a magnifying glass and a scrap of paper, to point out what happens to the retina in a similar situation.	*Handbook Safeguards*
	Serious eye damage, especially viewing through lenses, filters and pinholes.			*Handbook Safeguards*
microscopes	Microscopes using sunlight illumination are a particular risk	(Y7+)	Work near windows away from the sun.	*Handbook Safeguards*
Ultra-violet radiation				
short wave (315 nm or less)	Skin burning and permanent eye damage	T	The lamps have clear envelopes and need to be screened from direct view with an extra sheet of glass.	*Handbook Topics*
long wave	Headaches		The lamps have dark envelopes and rarely require extra screening.	*Safeguards*
X-rays	High doses may 'burn' skin	T	All staff must be trained.	*Handbook*
	Any exposure may cause cell damage	X	See Ionising radiation	

Table 18.4

Table 18.4   Gases and vapours under pressure and *in vacuo*

Equipment or Procedure	Hazard(s)	Suitability	Guidance Comments	Reference
Bell and bell jar	Disintegration (implosion) and scattering of broken glass	T	Check for cracks and scratches in glass which weaken it. The bell jar must stand on an approved glass or metal base. Use full safety screening and eye protection. Only glass vessels specifically intended for reduced pressure should be used.	
bell in evacuated flask		T	This has an added risk in that the evacuated flask needs to be handled and shaken.	*Handbook Safeguards*
		N	Thin glass 'bell covers' and flasks or window glass are not suitable.	
boiling and evaporation under reduced pressure	As above, with the added danger of vacuum pumps expelling hazardous vapour when unsuitable liquids are evaporated	N	Ethoxyethane (ether) should not be used with vacuum pumps.	
Boyle's Law apparatus				
mercury type	Spills of mercury	(Y11+)	Always use a containment tray.	*Hazcard Handbook*
pressure type	High pressure oil leaks and explosion	T	Follow manufacturer's instructions and always check for leaks.	
Cloud formation	Explosion hazard [pumping air into closed glass container and suddenly releasing pressure]	(Y10+)	Make sure pupils do not circumvent safety, for example by closing valve too tightly.	
Collapsing can				
using vacuum pump	Using cans which are not designed to collapse	T	Use thin-walled cans produced specifically for this demonstration whenever possible.	
using boiling and cooling	Possible hazardous residues in second-hand cans	T	Make sure cans are uncontaminated. Use safety screens and goggles.	
bromine				
'Guinea and feather'	Implosion of glass tube	T	All tubes should be checked for cracks and scratches because this seriously weakens them.	
Magdeburg hemispheres	Sudden separation of hemispheres when under tension	N (Y10+)	Particularly likely with small-diameter hemispheres. Close supervision is needed so that a sudden separation does not cause injury when the hemispheres are pulled apart.	
Mercury barometer	Spills of mercury when filling barometer tube	T	Use mercury tray and avoid skin contact; have a mercury spill kit available.	*Hazcard*
Steam engine models	Explosion of boiler	(Y7+)	Check the correct working of the safety valve before each use.	*Handbook*
	Fires spreading from liquid fuel		Use only solid fuel or liquified petroleum gas (LPG) and static engines.	*Handbook Safeguards*
		N	Ethanol should not be used.	*Topics*

18.2 References

1 Manual Handling Operations Regulations 1992

2 Petro/oxygen Explosion (Foundation Science Notes, p. 6) Bulletin 132, 1982. Petrol-Oxygen Explosion and Macpherson's Law (Safety Notes, p. 3) Bulletin 148, 1985, SSERC.

3 The Use of Ionising Radiation in Educational Establishments in England and Wales. AM 1/92, DFE.

4 The Use of Lasers in Schools etc. DES AM 7/70.

19 OUTSIDE THE LABORATORY

19.1 **General safety considerations**

When pupils are working or travelling off site, particular care must be taken to ensure that there are no significant risks to their health and safety. Occasions when pupils may be engaged in science activities, but not working in the school laboratory, include:

- their involvement in educational visits, for example to a museum;
- carrying out field work, either on short expeditions into the local environment during school time or on longer, often residential, field trips; and
- work-experience placements.

A short visit may not involve as many hazards as investigations in the field, but whenever students are off site, teachers must assess possible risks in advance and always take suitable precautions.

Table 19.1   Visits for field work

Source of Hazard(s)	Guidance	Reference
Planning	This is the key to the successful arrangement of all off-site activities. Many factors must be considered; the preparation required is governed by the nature of the work to be undertaken. Consultation with parents at an early stage and obtaining insurance, where appropriate, are essential. Employer's guidelines for out-of-school activities must be followed; among many considerations, these will often recommend or dictate the ratio of adults to pupils that must be adopted.	*Handbook*
Visits and journeys	Where the purpose of the out-of-school activity is merely to visit, for example a museum or to make a simple observation in the school grounds or immediate environment, the major requirements will be adequate supervision during transit/on site and consideration of risks from traffic. Provision for adequate hygiene will be important if handling objects in the outdoor environment is necessary. This is particularly relevant for visits to farms when animals may be handled, as there have been some instances of disease transmission where scrupulous hygiene was not observed.	*Topics* *Handbook*
Field trips	For more extensive outdoor work on a field course, perhaps at a specialist residential centre or even abroad, planning and preparation must be thorough. Of particular importance will be travel arrangements and ensuring that the facilities at a field centre and the expertise/qualifications of resident staff are appropriate for the planned activities. Leaders of groups of pupils must have adequate experience and may need additional, specialist training or qualifications, such as for first aid, in particular aspects of outdoor pursuits (for example if the activity involves sailing or walking on mountains) or driving a minibus. When driving parties of children, it may be necessary to share the driving and adequate rests should be taken on long journeys.	*Handbook*
Identifying hazards	Pupils must be adequately prepared for their work and have suitable clothing and footwear. They, and their teachers, should be familiar with relevant codes of conduct for work in the field (such as the Field Work, Mountain and Seashore Codes). The assessment of risk, which should form a significant part of the initial planning stage, will often require a preliminary visit to the site(s) used for the field work activities. Hazards, in addition to those of road and rail vehicles, may include water, hills, cliffs and steep slopes, holes (for example old mine shafts), falling rocks and, on urban 'waste' sites, unsafe ground, sharp masonry or broken glass and even pollution from hazardous substances.	*Handbook*
	For work involving fresh water, precautions against the hazards of Weil's disease, cyanobacteria (blue-green algae) and cryptosporidiosis must be taken. The water must not be drunk and open cuts should be covered. Small mammals and birds, dead or alive, must be handled with appropriate care (if at all). For practical work on geological sites, where there is a risk of rock chippings entering the eyes or of rocks falling from above, eye and/or head protection should be worn. Agricultural hazards, for example crop spraying, farm machinery and diseases caught through close contact with farm animals, may also be relevant.	*Handbook*

Table 19.2 Work-experience placements

Matter for consideration	Guidance	Reference
Arranging suitable work placements	When arranging for students to begin work experience, placement organisers must follow the guidelines and instructions that have been produced by their employers. It is also necessary to ensure that no work-experience placement is subject to a statutory prohibition or restriction. Circumstances where it is unlawful to employ young persons include working on ships, with lead and with dangerous machines. In addition, local by laws may restrict some work-experience placements. Work-placement organisers should ascertain from prospective placement providers which aspects of national and local legislation are applicable to the premises on which pupils will be working.	
Assessing the suitability of work placements	Placement organisers have the duty of ensuring that work experience positions offered by local firms etc. are entirely suitable and do not pose risks to the health and safety of pupils. Local guidance will indicate precisely what should be done in this assessment but will typically include provision for the following: • an initial visit to placement providers to check management systems for ensuring occupational health, safety and welfare. Those undertaking the visits should have a basic grounding in health and safety and access to specialist advice. This should include inspection of the written safety policy (if there are five or more employees), as well as the premises themselves. The nature of the work carried out on the premises of placement providers should have been notified to the local authority or the HSE; it is worth checking that this has been done. • confirmation that placement providers will ensure adequate induction and supervision of work-experience students. • confirmation that there are adequate procedures for dealing with accidents and emergencies and for first aid. • confirmation that there is adequate employer's liability insurance. • agreement on the appropriate number of hours of work per week (and their timing during the week) that the placement involves. • discussion of the work that the student will undertake and confirmation that any personal protective equipment required is available in the correct size and that training in its use will be given. In addition, any health-based limitations of the work should be identified, for example exposure to agents which would be inappropriate for an asthmatic, lifting operations which could aggravate a back conditions, etc. and the appropriate measures taken.	
Written agreements	It is good practice for placement organisers and placement providers to formalise, in writing, their agreement about work-experience schemes. In addition, parents or guardians of the work-experience students must be consulted and their written consent obtained before a placement is finalised.	
Policy for reporting accidents	If, despite all the precautions taken, an accident involving a student *does* occur in the work place, there should be an agreed policy for the action to be taken; this should also identify people to be notified.	

Table 19.2　Work-experience placements - *continued*

Matter for consideration	Guidance	Reference
Review and feedback	The procedures for placements should be subject to periodic review. This may involve visits to the establishment during and/or after a placement. If these visits involve other school staff, the latter should be properly briefed on relevant aspects of the placement. Feedback on the success of, or problems caused by, the work experience will be an essential part of this process and should involve all interested parties.	
Student support	Students embarking on a work placement should be suitably prepared for the experience. Briefing about their own responsibilities for health and safety, and what instruction and training they can expect from the placement provider, should be an important part of this preparation. Students should also be visited by placement organisers during their work experience and this can form part of the review of placement schemes.	

19.2　References

For addresses, see Useful Addresses.

1　LEA and Teachers' Associations guidance.

2　Safety in Outdoor Education, 1989. DFE.

3　Field Work Code. National Association of Field Studies Officers.

4　Mountain Code. British Mountaineering Council.

5　Seashore Code. Marine Conservation Society

6　LEA Work Experience Guidelines.

7　Work Experience: A Guide for Schools and Work Experi ence: A Guide for Employers, 1995, DFE.

8　Simply Safe 2: Guidelines on Basic Health and Safety at Work. Revised edition, 1991. Youth Education Service Publications.

9　The Work Experience Health and Safety Survival Kit, 1994. Right Track Productions.

10　Mind How You Go! 1992. IND(G)2. HSE.

PART C

Further Information

20 ABBREVIATIONS AND ACRONYMS

AM	Administrative Memorandum from the D/EE
ASE	Association for Science Education
ATL	Association of Teachers and Lecturers
Be Safe!	Be Safe! Some aspects of safety in school science and technology for Key tages 1 and 2. Published by the ASE (see Useful Publications)
BS	British Standard
BSE	Bovine spongiform encephalopathy
BS EN	a standard which is harmonised across Europe
BSI	British Standards Institution
CE(mark)	Communauté Européene (mark)
CHIP regulations	Chemicals (Hazards Information and Packing) Reg ulations 1994
CLEAPSS	Consortium of Local Education Authorities for the Provision of Science Services (support for schools and colleges is provided by the School Science Service based at Brunel University)
COSHH regulations	The Control of Substances Hazardous to Health Regulations 1994
CPL regulations	Classification, Packaging & Labelling Regulations (of various dates), now superseded by the CHIP regulations
CTC	City Technology College
DENI	Department of Education Northern Ireland
D/EE	Department for Education and Employment (formerly Department for Education (DFE) and Department of Education and Science (DES))
EHT	Extra high tension (electricity supply of 400 V to 600 V)
ELCB	Earth leakage circuit breaker
EU	European Union
F	Fume cupboard required
(F)	Although the use of a fume cupboard is desirable, small

	quantities of chemicals may be used in a well-ventilated laboratory
GCE	General Certificate of Education
GCSE	General Certificate of Secondary Education
GM	Grant-maintained
GNVQ	General National Vocational Qualification
Handbook	CLEAPSS Laboratory Handbook (see Useful Publications)
Hazcard	CLEAPSS Hazcards (see Useful Publications)
Haz Man	SSERC Hazardous Chemicals - a Manual for Schools and Colleges (see Useful Publications)
HFL	Highly flammable liquid
HMI	Her Majesty's Inspectorate/Inspector of schools
HMSO	Her Majesty's Stationery Office
HSC	Health and Safety Commission
HSE	Health and Safety Executive
HSWA	Health and Safety at Work etc Act 1974
HT	High tension (electricity supply between 25 V and 400 V)
IEC	International Electrotechnical Commission
IMS	Industrial methylated spirit
INSET	In-service education and training
IOB	Institute of Biology
IOP	Institute of Physics
ISBN	International Standard Book Number (a ten-number sequence which identifies books)
KS (1,2,3 or 4)	Key Stage of the National Curriculum
L (1,2 or 3)	Levels of work for microbiology and biotechnology in schools (see Microbiology: An HMI Guide for Schools, or Topics in Safety
LEA	Local education authority

LPG	Liquid petroleum gas
LTEL	Long-term exposure limit
LV	Low voltage (electricity supply, usually direct current up to 25 V)
M	mol dm-3 (a measure of concentration)
Micro	Microbiology: an HMI Guide for Schools and Further Education (see Useful Publications)
MISAC	Microbiology in Schools Advisory Committee
N	Chemical or procedure generally considered to be unsuitable or not recommended for use in schools
NAFSO	National Association of Field Studies Officers
NAHT	National Association of Head Teachers
NASUWT	National Association of Schoolmasters and Union of Women Teachers
NCBE	National Centre for Biotechnology Education
NIG	Education National Interest Group of the HSE
NRPB	National Radiological Protection Board
NUT	National Union of Teachers
NVQ	National Vocational Qualification
OEL	Occupational Exposure Limit
OES	Occupational Exposure Standard
OFSTED	Office for Standards in Education
OHMCI	Office of Her Majesty's Chief Inspector (more commonly called OFSTED in England, but not in Wales)
PAT	Professional Association of Teachers
PDSA	People's Dispensary for Sick Animals
PPE	Personal protective equipment
PPE regulations	Personal Protective Equipment at Work Regulations 1992
RCD	Residual current device

RIDDOR	Reporting of Injuries, Diseases and Dangerous Occurrences Regulations 1985
RPA	Radiation Protection Adviser
RPS	Radiation Protection Supervisor
RSC	Royal Society of Chemistry
RSPCA	Royal Society for the Protection of Animals
Safeguards	Safeguards in the School Laboratory, published by the ASE (see Useful Publications)
SAPS	Science and Plants for Schools
SCAA	Schools Curriculum and Assessment Authority
SI	Systeme Internationale (of units)
SSERC	Scottish Schools Equipment Research Centre
SSPCA	Scottish Society for the Prevention of Cruelty to Animals
STEL	Short-term exposure limit
T	Chemical or procedure suitable only for use or demonstration by teachers or technicians, or indicates concern about the access or involvement of pupils
Topics	Topics in Safety, published by the ASE (see Useful Publications)
TWA	Time-weighted average
WO	Welsh Office
X	Chemical or procedure banned nationally
Y (with number and + sign)	Chemical or procedure generally considered suitable for pupils in Years 7, 9 or 12 or above
(Y) (with number and + sign)	Chemical or procedure generally considered suitable for use by pupils in Year 7, 9 or 12 and above, but only under close supervision

21 USEFUL PUBLICATIONS

General safety

- Topics in Safety (ASE, 1988; ISBN 0 86357 104 2).
- Safeguards in the School Laboratory (ASE, 1988; ISBN 0 86357 083 6).
- Be Safe! (ASE, 1990; ISBN 086357 081 X).
- CLEAPSS Laboratory Handbook (CLEAPSS).

Health and safety in the workplace

- Guidance Note EH22 Ventilation of the Workplace (HSE Books, 1988; ISBN 0 11 885403 8).
- Guidance Note EH40 Occupational Exposure Limits (HSE Books, ISBN changes annually).
- The Responsibilities of School Governors for Health and Safety (HSE Books, 1992; ISBN 0 11 886337 1).
- BS 2092 Specifications for eye-protectors for industrial and non-industrial uses (BSI, 1987; ISBN 0 580 15858 6).
- BS 4163 Code of practice for health and safety in workshops of schools and similar establishments (BSI, 1984; ISBN 0 580 13866 6).

Pupils' safety

- Safety in outdoor education (DES, HMSO, 1989; ISBN 0 11 270690 8).
- Safety in Outdoor Activity Centres: DFE Circular 22/94 Safety in Outdoor Activity Centres: Guidance
- Work Experience: A Guide for Schools (DFE Publication)
- Work Experience: A Guide for Employers (DFE Publication)

Hazardous substances

- Amines: DFE Administrative Memorandum AM 3/70 Avoidance of Carcinogen ic Aromatic Amines in Schools and Other Educational Establishm ents.
- Asbestos: DFE Administrative Memorandum 3/86 The Use of Asbestos in Educational Establishments.
- Ionising radiation: DFE Administrative Memorandum AM 1/92. The Use of Ionising Radiations in Educational Establishments in England and Wales.
- COSHH: Guidance for Schools. Health and Safety Commission (HSE Books, ISBN 0 11 885511 5).
- Preparing COSHH Risk Assessments for Project Work in Schools. Available from SSERC.
- Hazcards: a set of A5 cards covering hazardous chemicals in schools. Can be used as a basis for risk assessments. Available from CLEAPSS.
- Hazardous chemicals: a Manual for Schools and Colleges (SSERC,

Longman. Currently being revised).

Diseases

- AIDS: DES Administrative Memorandum AM 2/86. Children at School and Problems Relating to AIDS.
- Asthma: HSE Guidance Note MS25 Medical Aspects of Occupational Asthma.
- Asthma: National Asthma Campaign School Pack (state whether primary or secondary pack required)
- Anaphylaxis: Anaphylaxis Campaign Guidance on the management of a child with anaphylaxis

Working with organisms

- DFE Administrative Memorandum AM 3/90 Animals and Plants in Schools: Legal Aspects.
- Microbiology: An HMI Guide for Schools and Further Education (HMSO, 1993; ISBN 0 11 250278 2).

Electrical safety

- Regulations for electrical installations (IEE, 1991; ISBN 0 852965 10 9).
- HSE Guidance Note GS23: Electrical safety in schools (Electricity at Work Regulations 1989) (HSE Books; ISBN 0 11 885426 7).
- HS(G) 13: Electrical testing (HSE Books; ISBN 0 11 883253 0).
- Building Bulletin 76: Maintenance of Electrical Services (DFE, HMSO; ISBN 0 11 270799 8).
- Maintaining portable electrical equipment in offices and other low-risk environments (HSE Books).
- BS EN 61010-1 Safety requirements for electrical equipment for measurement, control and laboratory use - General requirements (BSI, 1993; ISBN 0 580 22433 3).

Fuel gases

- IM/25: Guidance Notes on Gas Safety in Educational Establishments (British Gas, 1989).
- Guidance Note CS4: The Keeping of LPG in Cylinders and Similar Containers (HSE Books, 1986; ISBN 0 11 883539 4).
- HS(G)34: The Storage of LPG at Fixed Installations (HSE Books, 1987; ISBN 0 11 883908 X).
- British Standard 5482: Domestic Butane and Propane-burning Installations Part 1 (BSI, 1994; ISBN 0 580 22295 0).

Building and design

- Building Bulletin 7: Fire and the design of educational buildings (DES, HMSO, 1988; ISBN 0 11 270585 5).
- DFE Building Bulletin 80: Science Accommodation in Secondary Schools (DFE, HMSO, 1995; ISBN 0 11 270873 0).

- DFE Design Note 17: Guidelines for Environmental Design and Fuel Conservation in Educational Buildings (Architects and Building Branch, D*f*EE).

Fume cupboards

- DES Design Note 29: Fume Cupboards in Schools 1982 (DFEE; ISBN 0141-2825).

First aid

- First Aid in Educational Establishments (HSE Books, 1986; ISBN 0 11 883837 7).

Storage of Chemicals

- Storage and Use of Highly Flammable Liquids in Educational Establishments (HSE, 1986 leaflet).

Waste disposal

- Waste Management: The Duty of Care - A Code of Practice (Department of the Environment, HMSO, ISBN 0 11 752557).

22 Useful Addressess

Association for Science Education College Lane Hatfield Hertfordshire AL10 9AA..	Tel: 01707-267411
Association of Teachers and Lecturers 3 Northumberland St London WC2N 5DA.	Tel: 0171-930 6441
Anaphylaxis Campaign PO Box 149 Fleet Hampshire GU13 9XU	(please include SAE)
British Gas plc 326 High Holborn London WC1V 7PT	Tel: 0171-828 9722
British Mountaineering Council Crawford House Precinct Centre Booth Street East Manchester M13 9GH	Tel: 0161-273 5835
British Standards Institution 389 Chiswick High Road London W4 4AL CLEAPSS	Tel: 0181-996 9000
CLEAPSS **School Science Service** Brunel University Uxbridge UB8 3PH	Tel: 01895-251496
Countryside Council for Wales Plas Penrhos Ffordd Penrhos Bangor Gwynedd LL57 2LQ	
Department of Education **Northern Ireland** Rathgael House Balloo Road Bangor Co. Down BT19 2PR	Tel: 01247-279000

Department for Education and Employment Sanctuary Buildings Great Smith Street London SW1P 3BT	Tel: 0171-925 5000
D*f*EE Publications Centre PO Box 6927, London E3 3NZ	Tel: 0171-510 0150
English Nature Northminster House Peterborough PE1 1UA	Tel: 01733-340345
Her Majesty's Stationery Office PO Box 276 London SW8 5DT	Tel: 0171-873 0011
Health and Safety Commission Baynards House 1 Chepstow Place London W2 4TE	Tel: 0171-243 6000
Health and Safety Executive HSE Information Centre Broad Lane Sheffield S3 7HQ	Tel: 0114-289 2345
HSE Books **PO Box 1999,** Sudbury, Suffolk CO10 6FS.	Tel: 01787-881165
Institute of Biology 20-22 Queensbury Place London SW7 2DZ	Tel: 0171-581 8333
Institute of Physics 47 Belgrave Square London SW1X 8QX	Tel: 0171-235 6111
Institution of Electrical Engineers Publications from: PO Box 96 Stevenage Herts SG1 2SD	Tel: 01438-313311

Laboratory Animal Breeders Association Tel: 0116-252 2522
Secretary General
Unit of Biomedical Services
University of Leicester
PO Box 138
Leicester LE1 9HN

Marine Conservation Society   Tel: 01989-566017
9 Gloucester Road
Ross-on-Wye
Herefordshire HR9 5BU
**Microbiology in Schools
 Advisory Committee**
(see Institute of Biology)

National Association of   Tel: 01306-889070
Field Studies Officers
Membership Secretary
Juniper Hall Field Centre
Dorking
Surrey RH5 6DA

National Association of   Tel: 01444-416326
Head Teachers
1 Heath Square
Boltro Road
Haywards Heath
West Sussex RG16 1BL

National Association of Schoolmasters Tel: 0121-453 6150
and Union of Women Teachers
Hillscourt Education Centre
Rednal
Birmingham B45 8RS

National Asthma Campaign   Tel: 0171-226 2260
Providence House
Providence Place
London N1 0NT

National Centre for   Tel: 01734-873743
Biotechnology Education
Department of Microbiology
University of Reading
Whiteknights
PO Box 228
Reading RG6 2AJ

National Union of Teachers
Hamilton House
Mabledon Place
London WC1H 9BD

Tel: 0171-388 6191

Office for Standards in Education
Alexandra House
29-33 Kings Way
London WC2B 6SE

Tel: 0171-421 6800

Professional Association of Teachers
St James Court
Friargate
Derby DE1 1BT

Tel: 01332-3723372

Peoples' Dispensary for Sick Animals
PDSA House
Whitechapel Way
Priorslee
Telford TF2 9PQ

Tel: 01952-290999

Royal Society of Chemistry
Burlington House
Piccadilly
London W1V OBN

Tel: 0171-437 8656

Royal Society for the Prevention of
Cruelty to Animals
Education Department
Causeway
Horsham
West Sussex RG12 1HS

Tel: 01403-264181

Science and Plants for Schools
Homerton College
Cambridge
CB2 2PH

The Royal Botanical Garden
20A Inverleith Row
Edinburgh EH3 5LR

Tel: 0131-552 7171 ext. 465

Schools Curriculum and
Assessment Authority
Newcombe House
45 Notting Hill Gate
London W11 3JB

Tel: 0171-243 9273

Scottish Schools Equipment
Research Centre
24 Bernard Terrace
Edinburgh EH8 9NX

Tel: 0131-668 4421

Scottish Society for the Prevention
of Cruelty to Animals
19 Melville St
Edinburgh EH3 7PL

Tel: 0131-225 6418

Welsh Office
Cathays Park
Cardiff CF1 3NQ

Tel: 01222-825111

Youth Education Service Publications
Hebron House
Sion Road
Bedminster
Bristol
BS33BD

Tel: 0117-963 7634

23 Index

Abbreviations and acronyms are listed separately on pages 158 to 161

Printed in the United Kingdom for The Stationery Office
J69522 1/99 C10 10170